# *The Farm*

## LANELLE DICKINSON KEARNEY

**CAPPER PRESS**
Topeka, Kansas

Copyright © 1983 By LaNelle Dickinson Kearney
All rights reserved.
No part of this book may be used or reproduced in any manner whatsoever without written permission except in the case of brief quotations embodied in critical articles and reviews.

All characters and incidents in this book are fictitious. Any resemblance to actual people or events is purely coincidental.

———————◆———————

Published by Capper Press
616 Jefferson, Topeka, Kansas 66607

Cover Illustration and Calligraphy: *Catherine Seibel-Ledeker*
Production Manager: *Diana J. Edwardson*
Editor: *Michele R. Webb*
Proofreader: *Tammy R. Dodson*

ISBN 0-941678-29-6
First printing, October 1991

———————◆———————

Printed and bound in the United States of America

*The Farm was first published as a serialized novel in Capper's magazine (formerly Capper's Weekly) from March 29, 1983, through August 30, 1983.*

For more information about Capper Press titles
or to place an order, please call:
(Toll-free) 1-800-777-7171, extension 107, or (913) 295-1107.

## Capper Fireside Library

**F**eaturing the most popular novels previously published in *Capper's* magazine, as well as original novels by favorite *Capper's* authors, the *Capper Fireside Library* presents the best of fiction in quality softcover editions for the family library. Born out of the great popularity of *Capper's* serialized fiction, this series is for readers of all ages who love a good story. So curl up in a comfortable chair, flip the page, and let the storyteller whisk you away into the world of this novel from the *Capper Fireside Library.*

*Dedicated to Mama, to Jean,
and to Papa, who had such blue eyes.*

# Contents

CHAPTER ONE  *A Homestead in Kansas* ..........1

CHAPTER TWO  *Well Witchin'* ..........17

CHAPTER THREE  *Black Thunder* ..........36

CHAPTER FOUR  *The Raid* ..........50

CHAPTER FIVE  *Braving the Cruel Forces of Nature* ..........61

CHAPTER SIX  *Childbirth* ..........67

CHAPTER SEVEN  *Tinette O'Neil* ..........78

CHAPTER EIGHT  *Suitors* ..........88

CHAPTER NINE  *A Rocky Marriage* ..........106

CHAPTER TEN  *The Farm* ..........121

*The Farm*

CHAPTER ONE

## A Homestead in Kansas

**W**hen Calvin O'Neil and his bride Nancy first saw the piece of land on which they were to homestead, they were perched high on the seat of a covered wagon. That morning they had wearily jogged northward along the trail, up hill and down, as it led from Manhattan toward Marysville. About midway between these two Kansas towns lay the particular plot of land that lately had come into the possession of Calvin O'Neil.

While lying wounded in a makeshift Union hospital in Independence, Missouri, Calvin had heard a visitor describe his homestead. It was then that Calvin resolved to homestead in Kansas. The stranger said he staked a claim in 1858 and later registered under the Homestead Act. He cleared and broke sod on enough of the land to bring forth a fair yield of corn for a couple of seasons. He had taken the profit to build a one-room frame house that cost two hundred dollars to build, but now he was turning in his land grant and striking out for California. "It is a mighty good spot for someone," the man had declared.

The idea took hold of Calvin's fancy. He lay there in the hospital, his arm healing and his head dreaming about farming. He had always preferred farming to the dreary work in the railroad yards at Pittsburgh; and now that his father, Raither O'Neil, lay dead, a victim of the same Civil War battle in which Calvin had received his arm wound, there was no reason for him to go back to the railroad, or elsewhere. Calvin saw that each year life was slipping away from him at a seemingly faster rate. He was already thirty-one and had not a root to hold onto. Why not yield to the persuasion and assistance of this visitor, and to the cravings of his own heart, and head for Kansas?

Calvin's father had left Ireland in a huff over his brother's usurpation of the forge which should have gone to Raither himself. In the first place, Cloyd was a spoiled boy whose good looks as an adult only added to the trouble. In the same place, when Cloyd and Raither's father died, an O'Neil uncle took over the command of the forge and married a woman thirty years his junior. It was she, Anna, who favored Cloyd, who referred unnecessarily often to the handsome curve of his face, who threw every advantage his way, and who finally, after the sudden death of her husband, pushed Cloyd into the favored position at the forge.

Raither gave Cloyd one stunning punch in the gut and another one on that handsome face, allowed "Aunt" Anna one embrace, and left. He walked to cool off, and ended up walking onto a ship where he signed himself into servitude in exchange for passage to America. In Philadelphia he served his apprenticeship, of all places, in a forge. Every day was a reminder of things past.

When another slave of the system, a girl named Lily, became pregnant after her late night strolls with Raither, she hid from him but was finally forced to name him, truthfully, as the culprit. A man called Klatz, owner of the place where she was milkmaid, brought them together in marriage, and then dragged her back to the hovel where she had been hiding. Being, as he said, a religious man, Klatz gave Lily a Christian burial when she died in childbirth. Klatz named the baby boy Calvin, and he and his wife kept the child.

By the time Raither was free of his bondage, he had

more than his fill of a forge. He sought out the "religious" man who had his son, and offered to work to help pay for the boy's past keep. Father and son together labored on a farm, but their lives were virtually those of slaves.

Occasionally the pair would be sent into Philadelphia to market. Raither always returned with splendid profits as he was a born haggler with a sense of timeliness in the dealings. But he never was allowed to share in the profits. A place to stay, enough to eat, something to wear. That was all they had, with no prospect of change.

Raither did have ears, and once when in town he heard a man telling about his work on the railroads. After that, Raither took to questioning men about this type of work. In 1851 a railway reached a town called Pittsburgh, and Raither was told about this achievement and the industrial expansion that was bound to happen. Father and son did not bother to return to the farm that day. They left their carts and most of the money with a neighbor farmer and simply began walking westward.

The O'Neil men reached the "Smoky City" and both found work with the railroads. It was the forge again for Raither. He was made to wonder how Cloyd had survived.

Over a period of years, a letter he had finally written to Anna traveled across the ocean and back, but at last reached Anna in a little town north of Pittsburgh. Another couple of years passed and Anna's minister wrote that she and her second husband were well. She wanted Raither to know that they were working to bring Cloyd over. He had been ill and was no longer in charge at the forge.

It was time to move on. Anna and Cloyd were to be avoided, not to mention the war. Raither and Calvin worked their way to Nebraska. It was 1862, and generous federal land grants had been made to help the railroad companies connect to the west.

## THE FARM

It was during their stay in Nebraska that Raither and Calvin happened upon, and were pushed to join, a group of Union soldiers who seemed to be following their own direction. The northwest corner of Missouri was a conquered hotbed of Confederacy ideals, and raids back and forth into Kansas were still going on.

Somewhat south of Atchison, Kansas, Raither O'Neil was killed by an inadvertent shot in a misguided skirmish. A second shot sent streams of blood down Calvin's left arm. He lost consciousness, and later found himself in a Union cart with some wounded Missouri Grays who had been doing no more than protecting their property. He was taken to a makeshift hospital in Independence where gradually his strength returned. It was there that he resolved to homestead in Kansas. He even decided that he would let "Aunt" Anna know of Raither's death and his own general whereabouts.

Food was scarce, but at the hospital it was excellently prepared by a wisp of a girl named Nancy Hawkins. Sallow and fatigued, Nancy had arrived at the hospital late one afternoon with a large bundle of bedding and a pillowcase containing a few personal items. Not long before the war broke out, Nancy and her mother and father had settled on a tract of land edging the town of Independence, Missouri. They lived frugally, but because of the managing ability of Mrs. Hawkins, they lived comfortably. Through association with Union soldiers who camped near the Hawkins' property, Mr. Hawkins contracted malaria. He carried it into the little cabin. In only a short time both he and his wife were dead. Young Nancy packed some unused bedding, a dish or two, and her mother's ruby-red bowl, and moved to the temporary hospital in Independence. Before being taken over by federal troops, the lovely mansion had been the home of the wealthy Charles Fanforth family.

# A HOMESTEAD IN KANSAS

Except for her grief over the loss of her parents, Nancy was almost happy. She reveled in this fine abode, palatial to her. She felt that it was "home" to her, and she could have dedicated her life to its preservation. Then she met Calvin O'Neil, the best looking man she had ever seen. During the day she would try to leave the kitchen often in order to sit by and visit with him. He was older than she, and he seemed very comfortable with her attention. Calvin rested, ate, and talked. Then he moved about, collecting information from the other men, adding it all up in his mind. War was not for him and he must decide soon on his future.

So on May 7, 1863, Calvin O'Neil married Nancy Hawkins, and together they set out for Topeka, Kansas, to join a wagon train that was to be traveling close to their new homestead. Their wagon was a substantial one covered with white canvas. Through the top of the cloth ran a pipe which was connected to a cookstove set in the center of the wagon. Inside the wagon, Nancy stored bacon, beans, flour, salt, and a quantity of bread that she had made and dried. In the evening these articles would be set out on the bags of seed potatoes or on the top of the chickens, and sometimes even on the beehive, while a fire was started in the stove for Nancy to heat the milk and cook the bacon.

They ate outside the wagon as the inside was crowded with tools: a plow, spade, hoe, an axe, and a saw. The few cooking utensils, chairs, table, chest, and bed occupied the rest of the space. Under the driver's seat was a space where Calvin had pushed Nancy's trunk. It contained their bedding and a few personal items. Buckets and pails were hung under and behind the wagon to save space. At night Calvin would put the cream into one of the buckets, and the next day it would be converted into butter by the rocking motion of the wagon.

And so, alternately led by two oxen and then by two cows, the covered wagon carried Mr. and Mrs. Calvin O'Neil from Independence to Topeka, from Topeka to Manhattan, and from Manhattan, where they left the loosely-organized train, to their homestead.

The home which had served only as a hospital to Calvin was remembered by Nancy as a place of luxuriant beauty and warmth.

Each room had been outfitted with a solid wood floor and a large fireplace. Although the two grand rooms downstairs and the six rooms upstairs had been full of wounded soldiers, there was still a lovely dining hall, small sitting room, an enormous kitchen, the private quarters, and a general air of spaciousness for Nancy to revel in. She often found herself wishing that her parents could be there and see her housed in such a domain. The kitchen had been her special delight, equipped as it was with a great cookstove, wide fireplace, and every kind of cooking utensil. It was a real satisfaction to her to carry her hot soup and steaming cornbread into the cheery dining room where a blazing fire would be gleaming from the hearth.

But she had said goodbye to her life in that mansion, leaving it to take to the trail. And as the twenty-two-year-old bride observed the dugouts and soddies of countless pioneers the wagon train passed, her heart froze within her. Nancy looked up at Calvin on the seat beside her. He was an extremely handsome man with his large, round pansy-blue eyes and his tight black curls. Nancy knew she was a plain-looking girl. All her life she had been timid and quiet. Being beset by fears, her face appeared pinched and strained, her body stiffly shy and still. Calvin appeared not to notice, so she never disclosed her fears or the reason for them. Nevertheless, she worried in silence for fear that one of the hovels might turn out to be theirs.

## A HOMESTEAD IN KANSAS

A frame house might have been promised, but what actually awaited them could very well be a different thing. This was not so much a result of practical thinking on Nancy's part as a manifestation of her usual pessimism. She saw holes dug into the ground protected with covers of canvas or sheets, rooms dug into the side of a hill or a ravine with a few rails to make a door and occasionally a window; houses made of sod and plastered with mud; those were the dwellings she saw being used by people like herself and Calvin. Buffalo chips and prairie grass used for fuel; roofs leaked water and dirt; mud floors clod enough to leave the women with frostbitten feet. Sick at heart, Nancy yearned for her recent abode despite its hard work. She hardly dared look ahead.

The road from Manhattan to Marysville was better than most of the trails the young couple had traveled. They knew that about a mile north of the O'Neil property was a mill that a fellow by the name of Dan Pierson had opened in 1858. It was what was called a "combination plan," being used both as a sawmill and a grist mill. The location of this mill had greatly increased use of the road leading there, and some farmers were traveling as far as seventy miles to the mill. This thoroughfare lay to the west of Calvin's farmland by a good part of a mile. By veering to the east over a partially cleared path they found the trail that led northward past their property.

Perched high on the driver's seat of the oxen-led covered wagon, Calvin and Nancy O'Neil surveyed their land. Nancy, dressed in a high-necked, long sleeved, brown calico "sweeper" and sunbonnet to match, had thrown a blanket shawl around her thin shoulders and tried to withdraw into its warmth as she viewed the lonely countryside. Calvin, clad in heavy Army boots, hickory shirt and duck trousers, sat with his broad shoulders back as far as his stiff left arm would let them go, and drank in the sweetness of the view and the late afternoon air. Nancy sighed wearily for the dwelling she had lived in so contentedly while Calvin tried eagerly to locate the house that was to shelter them. Nancy wished for other years when her parents had been at her side while Calvin looked ahead to the day when sons and daughters would enrich the scene with their vigor and gaiety.

Calvin was not able to see the little one-room frame house from the hill where they had stopped, so he picked up the reins to urge the beasts forward. Then he laid the reins back down. With one hand he removed his cap and with the other one he ran his fingers through his closely set black curls. Then he turned and looked at Nancy.

"Nancy," he said. "I never was a fellow too good at religion. I might never mention it again. But there's somethin' I got to say."

His wife only nodded and stared straight ahead. Calvin swallowed, hit his left knee with his cap, and resumed speaking.

"We're startin' out with mighty little on our side. A few dollars. A wagon load of next to nothin'. No folks. Not even any friends or neighbors. No tellin' what'll come up — Indians, plagues. No use worryin' too much 'cause worryin' won't help. But when you're alone, you figger you got to have help from somewhere. Up there's the only place I know to ask for it." Without looking upward, Calvin pointed to the sky from which the bright rays of the sun had already begun to fade.

He waited more for a dying down of his own emotions than for a response from his wife and presently concluded, "If we make it through it'll be 'cause we asked for His help and got it."

Not being used to such talk from her husband, Nancy felt embarrassed. But in some way she also felt more at peace than she had for several days, and she slipped her small hand into one of his big ones. He squeezed her hand, placed the cap back over his curls, picked up the reins and bade the team to go on. Within a short while, still hand in hand, they drove up to their new home.

The former settler had cleared off quite a space near the center of the eastern end of the farm. A small trail led off the road and ended in front of a one-room dwelling. It was just as the prior owner had said, a frame house. Though it contained only the one room, it held the distinction of being the only frame house for miles around. Its rough native timber had been usable at Dan Pierson's primitive sawmill.

Calvin pulled the team to a halt just north of the house. They sat for a moment looking at their new home. There was no opening of any kind on this side of the structure. Calvin helped Nancy to the ground and she quickly started towards the house in search of a door, and "Lord help us," she thought, "windows."

"Here's a door to the west, Calvin," she called back, with her hand already on the latch to try it. The feel of the cold metal within her grasp made her withdraw her hand and look down to inspect the latch. Here on this roughly-hewn door was a metal latch of fine quality. The door itself, she observed, was held together with hardwood pins, but this hardware was attached with eight shiny screws. An odd bit of decoration for a pioneer house even though it was superior to most.

But what made Nancy O'Neil stand there quivering was not the latch, so wondrously out of place. The thing that made her stare in frightened bewilderment was a patch of dried blood. Whose bloody hand had touched the door? How long ago? What kind of signal was it? Was it an omen for them to leave without entering?

"Go on in, Nancy," Calvin called out in a hearty voice. "I'm bringin' your trunk."

Instinctively, and without being aware of moving, Nancy placed a hand on the latch and with very little pressure opened the door. It swung back, revealing a room about twenty feet long and eighteen feet wide. In the center of the south wall was a fireplace. On each side of the fireplace was a wide-paned window, the only glass windows in the area. Through their dusty translucence the remaining daylight entered. Under one of the windows lay a wild turkey, its bloody feathers staining the white ash floor. Close to the bird sat an old Indian, staring fiercely at the pale intruder.

Nancy, stunned, froze in the doorway, her frightened mind vaguely warned her not to turn her back on the burly, bronzed warrior whose bow and arrows lay on naked crossed legs. But to advance into the room, although it was her own home, was a feat she could not possibly accomplish. She felt the swift paralysis which had taken hold of her body; she could not even move her eyes away from the frightening spectacle of this savage clothed only in a filthy denim shirt that was too large.

The Indian had never seen a grown woman as small as this white woman, and he only stared back at her until there appeared behind her a solid, well-built man carrying a heavy trunk on his back. Fear had been evident in the woman's expression; but on the face of the big man was a determined look, as if to defy anyone to interfere with his plans. These characteristics were as obvious to Black Thunder as was their sex. And after sizing them up in this manner, he was content to sit and stare at them.

Calvin had heard many a tale about the Indians and the outrages they were capable of inflicting on a pioneer family that was widely separated from other whites. Immediately there came to his mind the recent tale of a young bride and her small brother who had been carried away from their cabin by a band of Indians. The boy's body had been found the next day, the scalp torn off his crushed head. The girl's body lay close to his, nude and full of arrows. This had been the result of a tribe's seeking revenge upon the young groom who had got them drunk in order to steal their furs.

Instantly it became clear to Calvin that he must not incur this fellow's anger. No telling how large a tribe might be camped nearby, although he knew that there were not many Indians left in this area. But feeling resentment against this intrusion and hot anger against any possible menace, he permitted himself to give one look of potential hostility straight at the semi-savage before pushing past Nancy and entering the room. The old Indian understood that here was a man to cope with.

Nancy, at last freed from her frozen stance when Calvin brushed past her, silently followed her husband in his every move. Together they went back outside leaving the door wide open. They unharnessed the oxen from the wagon and led them with the cows up to the side of the house. Making sure that the four beasts were tied securely to the trees by the house, the young couple began unloading their simple possessions. They guarded against any nocturnal theft by carrying everything inside the house. The table and chairs were placed near the center of the room. Along the north wall Calvin placed the rest of the furniture, bed touching chest, chest end to trunk. Unable to speak, her heart still in her throat, Nancy motioned to her husband to put the stove to the right as he entered. Along that same west wall, and also under one of the windows, they laid their dishes and foodstuffs.

Only once did the Indian move, and that was to incline his head enough to see Calvin place the coop of chickens and hive of bees in the corner behind him.

By this time dusk was at hand and the cows had to be milked. Along the way Calvin had always taken care of the milking while Nancy had prepared the meal. And, too, he had always considered her hands too tiny to be any good for milking. But on this night when not a word had been spoken aloud for over an hour and the air was heavy with silent apprehension, Nancy picked up a bucket and went to one of the cows as if milking were her most accustomed task.

When Calvin had milked his cow dry and his bucket full, he pulled his wife to her feet and squatted to finish her task. Hardly any milk remained and Nancy's pail was as full as his own. Straightening up he nudged Nancy with his elbow and relaxed into the grin that was so becoming to his round face and twinkling blue eyes.

"Strong hands, huh?" he teased.

But Nancy was not to be diverted from the feeling of panic which showed plainly on her face. "Shh," she warned, motioning with a nod of her head toward the house. The good-natured grin left Calvin's face, because it was indeed a serious circumstance to be facing the night with the big silent Indian encamped in front of their fireplace.

When Nancy and Calvin entered the house again, this time for the night, a fire blazed red in the fireplace. And on a horizontally hung spit above the fire the turkey was browning, its feathers and innards strewn on the floor where the Indian had been sitting. He had given up his former spot in favor of one closer to the fire where he could sit warmly and still turn the spit with a long stick.

The warmth of the fire made the O'Neils conscious of the chill that had come with sunset, and both of them would have enjoyed warming their backs at the open fire but neither dared to move that close. The smell of the roasting fowl made them realize how hungry they were and that they had not eaten for seven hours. Nancy thought of boiling some mush, but there was no water and she could not bring herself to speak, even about water. Mush could be cooked in milk, she was thinking; but not for anything, not even for a warm supper, would she presume to use that fire. There was nothing to do but drink the warm, fresh milk and lay out some cornbread and butter. And maybe some molasses. This she attended to quickly.

As Nancy cut the bread and poured out the molasses, she watched Calvin pour the milk. He poured a big mugful for Nancy and one for himself. Then taking a deep tin cup from the dish box, he poured out some of the sweet milk for the Indian.

Nancy put down tin plates for herself and for her husband, and Calvin laid one beside the Indian's cup of milk. Signaling for Nancy to sit beside him, he crossed over to the stranger and tapped him on the shoulder. The bronze warrior looked up, his face a mask as to any inner reaction. Still not speaking, Calvin pointed to the table indicating that the plate and cup opposite Nancy were for him. Without waiting for a sign of reply, Calvin left the fireplace and seated himself beside Nancy.

The Indian turned his gaze back to the fire. Presently he stood up, removed the bird from the spit and threw it on the hearth. From under the faded denim shirt he pulled a knife, and with several swift motions he cut two huge slices of rarely cooked turkey breast. Now he, in turn, stepped from fireplace to table and laid a mass of meat on the white people's table. Then he picked up the pitcher of molasses and poured some out on a piece of cornbread. Taking the bread in one hand and the cup of milk in the other he stepped lightly back to the fire where he sat down to eat his meal, replenishing it once with a large section of the turkey for himself.

Nancy had never eaten half-baked fowl before, and a few bites were quite enough. But Calvin, hungry after the trying experiences of the afternoon and evening, ate all of his and the rest of Nancy's portion. Twice the Indian got up to fill his cup with milk and again he covered his cornbread with more molasses. The first time he had licked the bread clean.

After the meal was over, Nancy wiped out the tin plates and mugs and set them on the stove to wait until morning for a more thorough cleansing. While she was ridding the table, the old brave got up and stretched lazily. Then relieving himself with a resounding belch he walked out of the door, leaving it wide open. The two remaining occupants of the one-room prairie house did not know whether to feel relieved of his departure or terrified at what might happen next.

Both of them felt as if they were being watched through the windows, and both of them kept the silence. Calvin hastened to pick up the remains of the turkey. Not wanting to throw them outside lest they should attract coyotes, he tossed them into the blazing fire. The forthcoming stench was revolting, and Nancy was glad for the widely swung door which neither had considered closing. The few logs which had been in the fireplace when they entered had now burned themselves into red embers, and the dying fire suggested bedtime.

Along the way, Nancy had seen the makeshift beds of most of the settlers. They had used anything from a "prairie feather" bed, consisting of several layers of hay, to a bed of poles. And so it was in the light of the fire and with some satisfaction that she helped Calvin assemble their honest-to-goodness bedstead. On top of the frame they placed a tick of prairie hay, one they had received on their way west in return for several loaves of the light bread Nancy had made and dried before leaving Independence. Next to the tick, Nancy proudly laid a featherbed. The young couple, in anticipation of their first comfortable night's sleep in several weeks, shook and smoothed the feather mattress until it was as soft as down could be and as smooth as the heavy ticking that enveloped it.

Going to the cherry wood chest, a wedding gift from the hospital-home in Independence, Nancy took out one of the four linen sheets which had been left unused by her parents and spread it on the bed. Her thoughts were sad. What would her mother think now, to see her in this lonely room forsaken by all but Indians, no telling how many? Little comfort these fine thick sheets were to a daughter who longed to see the dear hands that had made them. But then, Calvin was here and he was good to her.

At that moment her good husband was fumbling with a blanket that would probably be a little too much on this May night. There was a certain chill, but summer was close at hand and the night would not be too cool. Nevertheless, Nancy took the blanket from her husband's hands and tucked it neatly under the feather tick, and the bed was ready for use. Calvin quickly shed his clothes, and after making sure that his gun stood at the corner of the bed, he lay down on the soft mattress. His wife was just buttoning the crocheted collar of her billowy nightgown when she heard the door close. As nimbly as a deer she leapt over the footboard and threw herself close to Calvin. The two of them lay trembling in the outer portion of the big bed, leaving the half next to the wall untouched.

A dim light appeared in front of the fireplace. It was the light from a pipe. The Indian was back and he was smoking a pipe.

Hours passed with the eyes of the newly married couple steadfastly peering through the dark to where they had last seen the lone figure sitting cross-legged before the ash-filled fireplace. The dark and his absolute silence kept them from knowing that he lay asleep beside the hearth.

At long last Nancy heard the soft sounds of sleep coming from her husband. "Oh God," she prayed. "You heard what Cal said today. Now do it. Please."

But no sense of protection from Heaven or Earth came to the small figure that huddled close to her husband's warm body. A couple hundred minutes dragged by for Nancy before she, too, fell into a fatigue-born sleep.

CHAPTER TWO

## Well Witchin'

Morning dawned. It was to be the first really hot day of the season, and late at that. The O'Neils had felt the heat before awakening and had thrown back their blanket. Calvin felt warm and started to turn over but contact with Nancy's body awakened him. She opened her eyes at the same moment, and finding her face next to his, she gave him a rare, hasty kiss. His grin embarrassed her and she moved abruptly away. Right then the menace of the preceding evening came to her mind, and she sat upright. The Indian was gone and the door stood wide open.

This time Calvin closed the door and began to dress. The first chore he had to attend to was the business of finding some water. Not wanting to leave Nancy alone, he told her to help him carry the pails in search of a well or spring. They headed north because of the downward slope of the land. This piece of the property had never been cleared for cultivation and it was a beautiful sight. Bluegrass, slightly overrun, covered the ground. Here and there were clumps of shade trees.

At one spot a knoll of cottonwoods sloped gently to the west. The pioneer pair walked through this section hoping that its downward slope would lead them to a stream. But at its base was only a stand of high water grass. Veering back a little to the east, they continued north until they found their treasure. It was a curving stream, and on its bank, a few feet from the path Calvin had made, were two sunken barrels full of rain water. The delighted couple skimmed off the top dirt and dipped their pails into the precious water.

The walk down to the stream and back had to be made several times because the oxen and cattle needed water, and the chickens were squawking from thirst.

Until Calvin could cut a chimney through for the stovepipe, the fireplace was the only means Nancy had for cooking. Having had more milk and the remains of the hard cornbread for a snack in between water carryings, it was full noon before the O'Neils sat down to a warm meal. While Calvin had prepared suitable wood for the fire (he had promised Nancy to secure it only from a haphazard pile of logs south of the house where she could keep an eye on him from the window), his neat housekeeper combined some of the water with wood ashes to scrub the floor to its original whiteness.

For dinner Nancy laid strips of salt pork in her heavy iron skillet and put it on the fire to fry. Then she made coffee. It was really "coffee essence," a mixture made from parched rye and baked squash. As the makeshift coffee boiled, Nancy added a little flour and milk to the skillet to serve with the salt pork.

After stirring the gravy, Nancy started to straighten up and felt her flesh go cold. Nervous chills hit her spine. Her arms broke out in goose bumps. She stood as if pinned to the floor. Dazedly she stared into the fire until she felt Calvin beside her and saw him lean over to remove the skillet from the fire. Coming to herself a little then, Nancy reached out for the coffee kettle and followed Calvin over to the table. In the same spot where she had first seen him sat the Indian, and on the table lay a heap of fresh, wild strawberries.

The morning's work in the fresh air had invigorated Calvin. He stepped about the kitchen in a lively manner, talking for the first time in the presence of the Indian.

"Nan, you wash these berries and I'll spread out the table."

Commenting on how good the "vittles" looked, he dished out equal portions of pork and gravy in each of the three tin plates. Then he poured the steaming coffee into the three mugs and also over the pan of lightbread which was the last of the batch and hard as a rock. While Nancy set the bowl of berries and the pitcher of molasses down, Calvin motioned to the old Indian to get his food.

"Eat," Calvin explained.

The Indian replied with a short grunt and walked over to the table for his food. First, he took his plate of meat over to the fireplace where he squatted to eat. Then he took his plate back and filled it with strawberries and molasses. Twice he licked the sorghum off and went back for more. Finally, when the plate was empty he went back for his coffee, now cool, and his hunk of soaked bread. When his meal was finished he put his cup on the stone hearth and stretched out on the floor for a nap. From the table to the fireplace were spots of coffee that had dripped to the floor from the dirty hand that had carried the sopping bread. Now Nancy would have to use the water and wood ashes again. Furthermore, she would have to stay outside near Calvin, helping to clear out a grassy spot for the chickens.

Gradually, great changes took place on the farm. The very first job the newlyweds attended to was the cultivation of the land the previous settler had broken and used. With his gun strapped to his side, Calvin guided the plow over the rough sod. The oxen led the procession, then came the cows, and finally the plow with Nancy aloft its beam to handle the reins. Although it was summer when they began, Calvin and Nancy planted corn.

After turning the earth over in strips twenty to twenty-five inches wide, Nancy walked ahead dropping corn. Calvin followed, covering it with a hoe.

Working at this job from sunup to sundown, the eager farmer and his wife finally had the field northwest of the house cultivated. The team had made quite a nice road from the side yard to the field. And where the road met the corner of the field, Calvin built a sod pigpen for some strays he had found. Around the inside of the wall was a ditch to keep the pigs from rooting down the wall. The pigs could roam in the timber south of the pen to eat nuts, which so far would be their only food. They sought the shelter of the pen chiefly for the water they found in the log troughs there.

Several yards south of the house, Calvin cleared off enough space to make a chicken yard. At one edge of this clearing he found an acre of ground which apparently had been prepared for cultivation. Here he and his wife took spade and axe and loosened the ground sufficiently to plant a crop of potatoes. The weeds of the chicken yard soon grew thick because coyotes carried off the first and the last of the brood.

To the west of the house by about one hundred feet, Calvin fashioned out a sod dugout similar to the ones he had seen on the way West. Extending back into the slope of the ground, the front of the dugout reached about fifteen feet from the scooped-out dirt. A crude door of slats covered the entrance. Here the oxen and cows went for protection from storms. Here they would seek shelter in the coming winter. Just inside the slat door the Indian had laid some dried grass, and here he preferred to sleep. On stormy nights he occasionally sought the refuge of the O'Neil hearth, and also at mealtimes. All during the summer, all during his uninvited stay with his white brother, he contributed to the bounty of the table.

Together the three feasted on a variety of fish, wild fowl, and rabbits. Wild berries made up a good part of every day's diet. An occasional melon was brought in, from goodness knows where.

Life was extremely hard. Besides the long days spent out in the field, Nancy gave as many hours as Calvin in erecting the sod barn and the pigpen. Not only were the long hours tiring, but the extreme heat and terrific winds deadened their spirits and left them completely weary at bedtime. Calvin's stiff arm gave him considerable pain at night, and he tried to ease its throb by sleeping with it on a portion of the mattress that he had raised with his folded Army coat.

On Sundays they rested from their jobs of clearing and breaking the sod by doing the washing. Calvin would fill an old keg with hot water and soft soap and then his wife would dump the clothes into the hot mixture. After taking turns at prodding them with a strong stick, they would remove the articles one at a time. The clothes then were laid onto a clean rock and pounded with a mallet. Then they were strewn around the edges of the wagon to dry.

The O'Neils left the place only once that summer. That was one day around the middle of July when they harnessed the oxen to the wagon, now without the canvas covering, and went in search of the mill. Since it frequently happened that an empty house would be ransacked, Calvin had suggested that Nancy select a few of her favorite items to carry along. She filled her trunk with the linen sheets and the blankets, and the few bowls and kitchen utensils that had been saved from her parents' home. This load was too much for the trunk, so the dishes and utensils were laid in a nest of straw under the seat. On the floor of the wagon, beside the trunk, Calvin put his hand tools.

From what the stranger at Independence had told

them, they knew that the mill was located at a point north of the farm. They headed in that direction by way of the path which bounded their land on the east. About half a mile down the weed-covered trail, about where the O'Neil property ended, there rose some jagged rocks. At this point the weakening path veered to the right by several feet. Then it turned back north and made a steep descent down into the valley.

Just as the team started to feel their way down that rocky path, a band of Indian braves appeared, stealthily and silently, forming a straight line in front of the wagon. From each end of the line several warriors stepped out, clad only in loin skin breech cloths and feathered headdress. They climbed deftly aboard the wagon and removed the treasures that Nancy had taken along.

In a moment's time there was not a thing left in sight on the floor of the wagon. Without a word the red men motioned for the couple to get out of the wagon. Knowing of no way to combat such an army of thieves, Calvin jumped to the ground, lifting Nancy off the driver's seat, and he held her close to him. The piercing black eyes of one of the band spied the trunk, and in a second it, too, was handed down to a man who immediately disappeared with it into the cluster of rocks.

When the leader of the band turned to Calvin with the demand, "Eat," Nancy was able to move in the face of danger. Having looked ahead to the flour they would be able to buy at the mill, she had used the last of her stock to make blackberry pies which she had hoped to trade for calico or some other desired article. She had neatly stacked the pies in a little recess under the dashboard. In a flash she pointed to the pies with accompanying motions of eating. Tearing off the coarse towel wrapped around the pastries, the Indian saw the delicious looking pies. He made several explosive sounds of communication to his companions, then grabbed the pies and leaped to the ground. The entire group disappeared as quickly and as quietly as they had come.

The young couple took advantage of their foes' retreating backs and with much haste retook possession of the wagon and ordered the team of oxen to pull on. They did not really get control of their breathing until the wagon was once again on level ground. Then Nancy spoke quite unnecessarily in a whisper, "Cal, do you realize what they could have done?"

Calvin gave a relieved laugh as he answered, "Here I was a-feared you were worryin' about your sheets and nicknacks. But you hit it. We're lucky to have our scalps."

At the same instant Calvin held out his gun for Nancy to see and she touched her dress, where underneath the money bag still hung. Looking at one another's faces and realizing that the Indians had overlooked the white couple's chief necessities, they both broke into laughter. Nancy managed to say between giggles, "A gun and money and they took pies." At this remark Calvin gave her a serious look and said, "Well, them are goldurn good pies." They leaned against one another, with laughter that approached hysteria.

Later as they left the mill with seven dollars worth of flour and other staples, frightened and leery of more Indian encounters, the O'Neils decided to go home by the main road. Many wagons traveled long distances to reach Pierson's mill. Those from the south of town used this road which lay to the west of the O'Neil's farm.

Progress was faster on this more traveled trail, and by the middle of the afternoon Calvin realized that they were as far south as their farm extended. But there was no opening for them to get through to their land. The piece of property that lay between their land and the road was covered with a fine stand of corn.

"Nan," the man said, stopping the team and for a moment and using his hat as a fan. He ran his right hand through his thick, black curls, and finally said, "We'll have to go back and take the other road."

Nancy expressed her fright, and together they discussed other ways of reaching their home. But dusk could make a swift descent on an unfamiliar route. The only thing to do was to go back.

In order to relieve the tension and also because it was a subject that had come close to his heart's desire, Calvin spoke of the field they were passing.

"We need an opening to this road, Nancy," he declared. When she agreed, he talked more boldly and to the point. "Reckon that piece of land there is for sale?"

For this Nancy had no response. They possessed only a little more than three hundred dollars, and one night when she had wept from exhaustion, lonesomeness and womanness, Calvin had promised to use that savings to build her another room or two. Since her constant dream had its base in the memory of the mansion-hospital, the one-room frame house left her dissatisfied. Her husband's promise had kept her going, working at his side, day after weary day, using twice the strength that her ninety pounds seemed capable of giving. And now he wanted to use the money for more land. Well, she thought to herself, she already had one of the few frame houses in Marshall County, even though it was only one room. And she further reflected that she even had two windows. And a wood floor. And after all, it was money that Calvin had earned and saved.

Thus, long after they had passed the field, Nancy gained control of her lips and said, "Least you could do is find out."

Upon reaching Pierson's mill, Calvin turned the team to the east, and before long the oxen-led wagon reached the hill which would take them south. Calvin leapt to the ground to help lead the team up the steep incline and Nancy's deep-set eyes kept a steady lookout for the red men and their possible terrors. Her heart nearly stopped beating at the top of the hill. Just as the wagon turned to make the jog back west, the same line of Indians appeared on the path before them.

Instinctively Calvin jumped up to the seat to be near Nancy. As quickly as before, the red men, too, jumped up into the wagon bed, this time replacing each article they had stolen a few hours earlier. One brave slipped the trunk back under Nancy's feet, and another man handed her the pies which were by then covered with dirty fingerprints. Nancy was afraid of incurring his anger by smiling, but she did dare to hand back the pies as she blurted out "gift." Eagerly he snatched them and ran out of sight, the remainder of the band close at his heels.

Bewildered, the couple resumed their journey. As they neared their home, Nancy opened the trunk. There lay each sheet and blanket, apparently untouched. She turned to take inventory of the other articles and found each of Calvin's tools and all of her kitchen utensils. Only one item was missing: the ruby-red bowl that had belonged to her mother.

When they got home, Calvin's body ached as if he had done a day's plowing and Nancy's head felt twice its size and about to burst. Feeling so wretched, they decided to get the milking done and then settle for a simple supper of coffee, milk, and mush. Along with the purchase of the usual "coffee essence," Calvin had also bought one sack of "preferred coffee," a mixture of two parts peas and one part coffee. Many people considered this mixture stronger and more flavorful than real coffee, and the O'Neils agreed that their difficult day warranted the use of it. Putting the kettle on to boil, the two of them went out to milk.

# THE FARM

When they had finished their chore and returned to the house, there was a small orange melon on the table. It lay in the ruby-red bowl that had belonged to Nancy's mother. In front of the fireplace with his back to his friends sat the silent old Indian.

Two-and-a-half months went by. Then to the O'Neils came a great day of joy. On the morning of August 21, 1863, the sun shone brightly over the farm as if it knew that certain pleasures would come to them before it would hide its face behind the western horizon.

Soon after Calvin had taken the trip to the mill, he had decided to search out the owners of the adjoining field and see about making it his. After all, Nancy had seen the reasonableness of it and had given her consent. Remembering the friendship the Indian had shown them on the day of the Indian raid on their wagon, young O'Neil returned the compliment with a display of confidence. He briefly stated his plan to his wife and without waiting for a response, acted upon it.

At noon he left the dinner table and picked up his rifle, making preparation to leave the house. Before going, he stepped over to the old man and motioned for the warrior to follow him outside. The Indian followed. Outside of the open door, at which Nancy had hung a piece of the wagon canvas in lieu of any netting, Calvin turned to his companion and spoke in clear English, "I am goin' to walk across the fields yonder." To emphasize his meaning he pointed to the west and tromped up and down as if he were walking. Then signaling for the old man to sit down, he added, "Watch." The Indian raised his right hand before his face, an action interpreted by Calvin as assent. The white man raised his hand in like position, then turned and walked toward the west.

First, he walked across what Nancy called her side

yard. It abruptly became, without definition, the wagon lot. At the far end of this lot stood the dugout barn. Upon reaching the crude structure, he turned left and walked south as far as the pigpen. A little to the right of the sty stretched the cornfield. Calvin walked through his field, noting that it did not look too bad considering when it had been planted. When he reached the eighty acres that lay to the west, however, he realized that here indeed was a superior crop of corn.

In quick calculation he believed that his own crop would only yield about half of his neighbor's crop. Calvin stood his rifle beside himself and removed his cap. He gazed longingly at the land while slowly running a muscular hand through his close-knit curls.

"Jehoshaphat," he said aloud. "I'd give all the money I have for this strip of land."

It was not only the rich yield that appealed so much to the farmer, and not only the fact that it would give him an opening to the main thoroughfare. The longing also arose from a desire to own the entire section. He wanted all of the land, not an incomplete part. Almost like an artist who feels the need to finish the canvas, Calvin sensed that the beauty of the land was marred by not presenting a complete picture.

Once more donning his cap, he picked up the rifle and hurried through the field to the road. Across the road he found the owner of the property, a man called Hod Beach. Beach had purchased the land for a few dollars, and with it the land that formed the southern and western boundaries of the O'Neil farm. It was rumored that he lived there in an old shack with an Indian girl whom he had bought for a bag of bright beads. He was convincingly reported to be a successful farmer, but on this day he was lounging with two other men on the porch of a cabin that faced the road where Calvin was standing. Half drunk with strong whiskey, Beach felt suspicious and vaguely worried at seeing a stranger coming from his property. When he tried to reach for his gun, his host, well aware of Beach's condition, had already quietly removed the weapon.

By the time Calvin reached the cabin, Hod Beach was staggering to his feet hurling oaths as to the purpose of the intrusion on his property. When Calvin mentioned his intent, anger rose so swiftly in Hod Beach's heart that he could barely sputter out his refusal. Forgetting his gun, Hod picked up his whiskey jug and unsteadily made his way down the road cursing and yelling back to Calvin never to meddle in his business again.

Another man came out of the cabin to ask the men what the trouble was. One of the men on the porch answered, "Hod's a-feared someone's goin' to find out about his swindle and take his claim away." At that remark the other men on the porch laughed heartily, calling out to Calvin who had started to walk back across the road, "You a neighbor?"

"Yup, over east a ways," Calvin turned back to answer. "The next property. O'Neil's the name."

"Ed Winchall's mine," the heavy-set man said, "and this here's Royal Lee, best water witch round about."

As the men shook hands, Calvin's thoughts took hold of the words "water witch" and began to piece them in his mind. He and Nancy had been carrying water from the sunken barrels at the little stream ever since their arrival. Now that he could not buy the land, he would still have the money. Why not fix up a well close to the house, he thought.

"What's yer well-witchin' worth?" Calvin asked the professional man.

Royal Lee squinted his eyes and asked, "Have you got ten dollars?"

## WELL WITCHIN'

"Jehoshaphat, for well witchin?" Calvin exploded.

"That an' drivin' a well," Lee replied with an accomplished spit of tobacco juice for emphasis.

On their trip west the O'Neils had seen a well "put down." Two neighbors worked together using a rope, log windlass, and iron-bound buckets. But to "drive" a well was a different matter, and Calvin understood that it required more tools and a keen knowledge to effect the job.

"How're you aimin' to do it?" Calvin asked.

Royal Lee described the scheme used by a company at Marysville, some twenty miles north. The idea was to drive a sharp-pointed pipe into the ground. Lengths would be added until the point hit water-filled sand.

Calvin listened seriously to Royal Lee's plans, then thoughtfully turned the idea over in his mind. Finally he asked, "When could you be at it?"

Royal Lee named a date when the men were to be in Calvin's vicinity. He said, "The day before they're to drive the well, I'll be over to witch fer water."

With the arrangements made, Calvin started to leave. Ed Winchall thought to ask if he had family, to which Calvin replied that he had a wife.

"Stayin' alone?" the neighbor queried. In response to Calvin's nod, Winchall added, "Lots of Indians around here right now again. Tribe of Kaws, most of 'em killed off by the Sioux when Lean Dog camped close. But plenty around to do damage."

Calvin thought of the old warrior at home and asked Winchall, "Did you ever see an old cuss in a blue shirt roamin' around?"

"Gawd sakes, man," Winchall said, "that's old Black Thunder, meanest skunk of them all. He's claimed more scalps than Lean Dog. Hod Beach's squaw says old Black Thunder's plannin' a war raid on us settlers 'fore long."

Calvin O'Neil knew real fright as he hastened towards

home. Why had he ever dared to leave Nancy at home? And alone. No, it was worse than alone. He had left her with a savage. Why had he trusted that Indian? All the tales he had ever heard about Indian terrors against white women rose into his head. He became so tortured with his fears that he broke into a wild run. Upon reaching the pigpen, he thought, "Now what if all is well?" He reasoned that if he went running in, the old Indian would realize his doubts. The new relationship might easily end. "Just in case," Calvin said to himself, "I'll slow up."

As he cut across the barn lot he was almost afraid to look toward the house. But when he stepped into Nancy's side yard he saw Black Thunder sitting just outside the door. Both his arms and legs were crossed, and he sat as erect and businesslike as if he were conducting an important council. Upon seeing his white brother he stood up, held his right hand up in front of his face, and spoke the first English Calvin had heard him say, "Good." Then he walked swiftly across the lot and out of sight.

When the relieved young man entered the house, he was surprised to find Nancy lying across the bed, crying. Fearing that she had been molested after all, her husband rushed to hear her account of the afternoon. He was astonished to find that her tears were the result of anger, anger aimed at him. It was unusual to hear his wife speak so long at once and with such vehemence. Calvin was really shaken by her feelings. It seemed that she did not mean much to her husband since he could leave her alone with a savage for protection. He probably took the cows with him to be sure of their safety, but a wife did not matter. Why had she ever left her place of shelter? Why, oh yes, why did her parents, the only ones who had really loved her, have to die?

Calvin was really more affectionate of nature than his wife allowed him to show, and being used to hiding his

## WELL WITCHIN'

natural inclinations towards holding Nancy, he tried to comfort her by other means. He told her in glowing terms of the water well they were soon to have, but this did little towards helping the situation. It was then that he reissued his former promise to add more rooms by winter. At hearing this, Nancy sat straight up and snapped, "How will you pay for it? Sell the land you just bought?"

Calvin then described to her his meeting with Hod Beach, and explained how hopeless it was to try to buy the field as long as Beach held it.

Realizing that Calvin's hopes had been crushed and that her greatest dream was about to come true, Nancy O'Neil once more threw her small body across the bed and wept for more reasons than she could name.

On August 20, 1863, Royal Lee officiously strode into the O'Neil's house about noontime. Nancy had been frying rabbit and boiling green roasting ears. The water witch could hardly keep his eyes off the stove and his nose heartily sniffed the delicious aroma. In those days company was welcomed with enthusiasm, so Calvin immediately set out one more tin plate, mug, and fork.

Black Thunder had been sitting on the cold hearth waiting for his food. But on seeing Royal Lee enter and prepare to stay, Black Thunder scowled, got to his feet, and walked across the room to the door. There he stopped and stood with arms folded, staring at the heavy canvas. Nancy filled his plate and cup and handed them to the old man, motioning for him to be seated. He shook his head gravely and took the food just outside the door where he sat to eat. Upon finishing, he pushed the plate and cup under the canvas and left sullenly.

Royal Lee watched all this in astonished silence, and after a hasty meal, he immediately began his work as if he could not get out of there too soon.

Taking up a forked willow stick, the only implement

he had, he pressed his hands firmly on its ends. Then with his hands against his knees, he bent low and walked back and forth over every inch of ground in the side yard. No activity came from the stick. Calvin said then that they would try the spot south of the house and west into the barn lot. Frequently, Lee would stop to straighten up and stretch his back. He would lay the stick down, take a black rag from the pocket of his dirty denim pants and wipe his face and neck free of the perspiration acquired by his strenuous task.

Together he and Calvin walked over many feet of ground, stepping quite slowly so as not to miss any possible pull on the stick. They were beginning to despair. The spot the O'Neils had wanted for the well was west of the house on the edge of the barn lot. But Royal Lee shook his head as there was plainly no water there.

Calvin suggested going to the house for a drink of water, and as the two men headed back in that direction they heard the whir of an arrow flying through the air. It seemed to have come from north of the house, and its point dug into the ground at a spot some fifty feet north of the house and about equal with the unmarked division of the side yard and barn lot. Royal Lee swore and spat tobacco juice, while Calvin ran to the corner of the house. He peeked around it in search of the arrow sender but no one was in sight. Then he sped to the back of the house until he could see all of the ground that lay on the east as far as the little trail that bordered the farm. Not seeing a sign of any intruder he circled back to the north until he reached the arrow itself. Lee sauntered over to him, the stick dragging from one hand.

"What's the meanin' of this — " he began when suddenly the stick seemed to be twitching. Grasping it firmly in both hands, with his hands against his knees, he resumed his stooped posture and started walking slowly around the arrow. As he passed over a spot which was at a slight angle southeast of the missile, the stick began to move and finally the main stem bent over to the ground. The willow rod indicated a vein of water.

Water witch Lee grinned and grabbed Calvin's hand. Calvin gave a whoop as they had a hearty handshake, and finally tossed his hat high into the air. Lee promised to return early the next morning, August twenty-first, with the well-digging crew.

Calvin was jubilant, and after saying goodbye he turned towards the house to call Nancy to come join him. She had been busily cooking in anticipation of the well crew she was expecting on the morrow. Not knowing how many to prepare for, she had decided to make five pies, using the precious flour and the gooseberries that she and Calvin had picked from wild bushes. On hearing her name called she guessed that the water witch had found a vein. She wiped her hands on her apron and stepped outside to look for the men. Knowing where they had planned to dig the well, she was surprised to see Calvin standing north of the house, down the slight slope that led away from the house.

"Here's where it'll be, Nan," he said with the good-natured grin which gave him such an air of lightheartedness. He stood there bareheaded, his black curls glistening from the heat. His eyes, wondrously blue and enormously round, sparkled with pleasure. He was a very handsome man, while his wife's rather obscure little face was extremely plain and usually marked with fret lines.

"Are you sure there will be water there?" Nancy asked as she neared the spot.

"Sure," he said. "See that arrow? Shot down into the ground but it hit water and bounced back up."

"Oh, Calvin," she said impatiently, not wanting to hear

jokes from her husband. "What arrow is that?"

"I told you and that's all I know to tell you." Taking her hand he led her back to the house saying that tomorrow was to be a great day.

The next morning Calvin and Nancy were up and about their tasks before daylight. As soon as Calvin could see, he went out in search of prairie chickens. As they were still at roost here and there, the job was easily finished, and about five o'clock on this clear day, he walked into the kitchen with several of the little dead birds. His chore was to shoot them, Nancy's to de-feather, de-entrail, wash, cut up, and fry them. First, however, she needed to fix a bit of breakfast for the two of them and attend to another chore or two. Before they had finished the milking, the wagon load of workers and their equipment rolled onto the O'Neil property. It was a businesslike crew, and with Royal Lee to show them the spot, they were able to get right to work. They began by driving the sharp pointed pipe into the ground. Lee showed them the spot where the famous arrow of the day before had lodged. The men shook their heads in bewilderment that such a thing could have happened and the water witch still be alive to tell about it.

On the drive west, Calvin and Nancy had seen many open-faced wells which the pioneers used by letting buckets down to be filled. Sometimes these deeply dug wells became death traps as passersby would stray onto them at night. But it was as satisfactory an arrangement as the O'Neils had expected. So Calvin stood completely awestricken when the men carried a pump out of their wagon and began to place it above the pipe. It took a little bit of doing to get it fitted onto the pipe properly and lower it until it touched the ground. Behind Calvin's amazed eyes, his brain was working. He was thinking that when he had heard the asking price of ten dollars, it had seemed like far too much money to put out. But it had not occurred to him that even such a sum could buy this great luxury. And he was thinking that his farm was getting to be as good as any they had passed by on their travel. Not only as good, but better!

Calvin had the distinct feeling that Lee's account of Black Thunder's presence in the O'Neil house, and of the arrow too, had something to do with the haste in which the men worked. And when they were through with the work after noontime and left without the meal which was their due, he knew good and well it did. Nancy had the chickens and potatoes cooked and was preparing the gravy when she heard the wagon leave. She ran out to find the reason for the men's early departure.

Calvin was on his way to get her and show off the new prize. The sting of her disappointment at not having her bounteous meal eaten and praised was considerably sweetened by the sight of the pump.

Calvin and Nancy were like little children in their excitement. Each one had to pump several bucketfuls of the water, and each declared that it was the tastiest, most satisfying water they had yet drunk.

CHAPTER THREE

## Black Thunder

In the midst of the excitement, the O'Neils cup of joy began to overflow with the arrival of a new and unexpected pleasure. They had stopped to give themselves and the pump a rest when Calvin thought he heard the sound of grinding wheels. Hardly able to believe his ears, he turned to face the sound. Turning into the O'Neils roadway was a covered wagon.

"Why, Nancy," Calvin cried, "we've got company!" Then breaking into a run he went with outstretched hands to greet the stranger who was leading the ox team.

"Yes," Nancy murmured half aloud and yet to herself, "our first real company."

The wagon had drawn to a stop close to the pump, and Nancy's eyes were eagerly searching over it for the sight of a woman. Months had passed since she had visited with a woman, but until this moment young Mrs. O'Neil had not realized how desperately she had been wanting a chance to be near someone of her own sex. Someone like herself, weak in the face of danger, sick for the niceties of a more settled community, anxious about the future.

"Nancy," she heard Calvin saying, "this here's Julius Katchall. He's homesteadin' up north a ways."

"Ya," answered the newcomer, "we must be neighbors."

"Neighbors," repeated Nancy, her eyes shining with hopes of friendly people living close to them. But her face was set with little lines of worry lest it should not be true.

Calvin slowly ran a hand over his downcast head, and sensing his wife's thoughts said, "Well, not real close neighbors. They missed their road today. They're to be considerable 'tother side of the mill."

Nancy, realizing the actual distance that was to lie between them, sighed and seemed about to cry when she felt a soft hand on her arm. She looked around and there at her side stood a woman, not much taller than she but almost twice as wide. Her hair was dirty and greasy from the long trip and its heavy braids encircled the woman's head. Her exposed ears were large and in each lobe was a ruby earring. One tooth was missing from each side of her mouth and as she smiled at Nancy those two gaps seemed enormous. But to the young girl who had been in need of a woman's companionship, Henrietta seemed like an angel. Complete strangers, as yet unknown to one another, they embraced, laughing and weeping, and shaking their heads in disbelief. This coming together was beyond their understanding.

By the time they learned one another's name, the rest of the Katchall family was out of the wagon. There was Jule, a boy of seven, and Adolph, barely six. Clinging by now to her mother's long, black, calico skirt was a shy little girl of about four. Her name was Clara.

When Henrietta had pronounced the names of her three children, it seemed she had said nearly all the words she knew that Nancy could understand. From then on she would either stammer out a broken sentence or rely upon Julius for interpretation of her German words. The children spoke entirely in German, but their father proudly declared that they knew English as well as anyone. They kept this accomplishment hidden.

The O'Neils led their new acquaintances over to the pump and bade them taste the fresh water.

"Ja, ja, das ist gut," Henrietta said with real feeling. It

had been several days since she had tasted such good water, and so clean.

"Dis pump," questioned Julius, "how it come here?"

So while Calvin explained about the water witch, the ten dollars, and the pump company, his wife took the plump German woman to the one-room house. Henrietta was impressed by the frame structure and the metal latch on the door. She stepped lightly onto the wood floor as if she were testing the strength of the white ash boards. Upon seeing the fireplace with its stone hearth she forgot about the flooring and began an inspection of the hearth. This delight gave way to the surprise of finding two windows with glass panes. In fact, Henrietta took a tour of the room which lasted several minutes. She saw everything and her comments were brief although complimentary: "Gut. Schoen. Ja wohl."

During this time Nancy had been finishing up the meal that she had originally planned for the welldiggers. How happy she was at this moment! As happy as minute Nancy Hawkins O'Neil had been for some time. She was entertaining in her own home, and effecting it with a fine display of food. She and Henrietta filled the plates for the men and boys, and watched them sit down to big helpings of chicken, potatoes, gravy, and cheese. Little Clara stayed close to her mother, although now and then she chanced a look and half-smile at Nancy.

While Nancy was refilling the men's mugs with hot coffee, she heard a light scraping at the door.

"Black Thunder," she said. "I forgot all about him." She set the big coffeepot back onto the stove and went to the door to look for the Indian. There he stood, perfectly erect with arms akimbo, facing the canvas covering. Nancy peeked out and motioned for him to come in. Black Thunder raised his head high but remained silent.

Nancy was so thoroughly elated with her guests that

she risked speaking directly to the warrior.

"Friend," she said as she pointed behind her. Then "friend" she said again, this time pointing to the Indian himself.

His face registered understanding but he refused to move. One arm unbent from the other and came straight up before his face. Then he slowly turned around and sat cross-legged in front of the door. Nancy took his gestures to mean friendship and went to fill him a plate. By the time the men and boys had finished their pie, Black Thunder had eaten his food and pushed his plate just inside the door.

The women then sat down to eat, placing little Clara between them.

"Your baby," Nancy said as she patted the little girl's small round head.

"Nein," Henrietta answered softly, "baby die."

"Oh," Nancy spoke sympathetically, "you lost your baby?"

"Baby die," the mother repeated with a hint of a sob.

Clara looked up at her mother, realized the cause of her mother's tears, patted her hand and said, "Tinette."

"Tinette?" asked Nancy, eager to have the sorrowing woman talk about it.

Henrietta nodded. "Ja. Tinette. Tinette mein friend zu heime; freund — friend von Sveeserlandt."

"Oh, Tinette was a friend from Switzerland."

"Ja. Und Tinette meine baby."

"You named your baby Tinette. That is a pretty name."

"Schoen, Tinette. Die in wagon." With these last words tears broke free and flowed down Henrietta's face. Nancy realized that the baby had died on the trail and more than likely had been left in a lonely grave on some desolate hill.

It took a while for Henrietta to quiet herself, and in the

process she and Nancy became close friends. Not much was needed to seal the bond. An understanding nod let the bereaved woman know that her friend also knew the bitterness of death, the slight pressure of Nancy's thin arm around Henrietta's broad shoulders, and an encouraging smile to help lighten the load.

Clara finished her last sip of milk and as if on cue the men appeared at the doorway.

"Come, Henrietta," said Julius Katchall.

Nancy stood up in surprise and asked, "What? You mean to be gone already?"

Calvin had felt that his wife would expect to keep her company overnight, and in anticipation of her disappointment he had arranged a visiting day not far in the future.

"Julius says that they must make it on before dark, Nan," he explained, "but he says for us to pay 'em a visiting day not far in the future."

"How would we know where to go?" Nancy asked doubtfully.

"Says to give him four weeks to the day and he'll be at Pierson's mill bright and early in the morning."

A month could stretch out like a year when all there was to see were menfolks, but this dim promise of a future visit was all Nancy had to cling to. She turned to her new friend with as cheerful a face as she could manage and tried to relay the news. With Julius' help Henrietta understood, and the women walked to the wagon arm in arm. Both women were homesick for what they looked back to as home, and both hearts were full of their newly found friendship. Tears came easily.

As Nancy saw Henrietta lift Clara into the wagon, then disappear behind her, her courage failed completely. She turned and ran for the house, crying aloud. As she pushed the canvas aside, she saw the last gooseberry pie and recognized it as a parting gift for her friend. She grabbed the pie and turned to leave the house, coming face-to-face with Black Thunder. It was the first time he had seen her cry, and he looked at her tear-stained face with amazement. Once more she spoke the word "friend" to him and ran towards the moving wagon. Calvin saw her coming and shouted for Julius to wait. Taking the pie from Nancy, he ran to the wagon and handed the gift up to Henrietta.

Together the O'Neils walked out to the trail, waving to the rolling wagon. Somewhere, as if from out of nowhere, a lone figure appeared and walked at a short distance behind, yet out of sight of the wagon. It was Black Thunder, ready to protect his friend's friend.

As Nancy O'Neil watched her new-found friend disappear into the distance, loneliness took hold of her to such an extent that there did not seem to be air to breathe. She ran back to the house and threw herself onto the bed in a frenzy of weeping that was developing into hysteria.

The bewildered husband took a seat beside her and tried all he knew to soothe her. No stroking of her head or soft words had any effect on the distraught nerves. She refused to let him hold her hand, pulling both of them underneath her little body as she lay disconsolate with her face buried in the pillow.

Seeking to divert her, Calvin said, "One of these days we'll have a family of our own. Those Katchalls have a fine family. Boys are a great comfort to people. Lots of boys needed out here on this new land, too." The pauses were becoming awkward but he tried once more, "Did you notice how satisfied Julius Katchall was with his boys?"

This time there was no pause and it became clear to Calvin that he had opened the wrong subject when he saw his wife sit up with a suddenness that startled him.

"All you care about is having boys. Boys! Boys!" she yelled out amid sobs. "A woman needs a girl. A woman who has lost her parents and has no friends or kin close by to visit and call upon for help, she needs a girl." She fell back onto the pillow and beat her fists on the bed as she spoke in a determined tone, "But I don't want a girl or a boy. I have no place for a family. I don't have room for a family. Even in my parents' old cabin, we each had a room. I don't even have a decent home."

As the evening wore on, it became obvious that the reason for Nancy's spell of nerves lay not in any hoped-for pregnancy, but in quite the reverse. Her physical misery and emotional upset kept mounting until Calvin was ready to say or do anything that would serve as a sedative. This "sleep potion" came in the form of a promise to begin to enlarge the house as soon as possible.

During the fall months that followed, Calvin put all the time he could spare on the house. First, he purchased lumber from Pierson. As cash became low Calvin took to cutting his own logs and hauling them to the mill. Pierson kept half of the lumber and Calvin carted the rest home. Because of the O'Neil's heavy patronage, the mill operator forfeited his right to the slabs. This was the waste part of the tree that was fit only for firewood. While Calvin worked on the rooms, Nancy corded the slabs into neat rows for winter use.

In the north wall Calvin cut two doors. The east one opened into a room which Nancy hoped to furnish as a sitting room, the other one opened to a bedroom. The fireplace in each room was against the center wall and had a common chimney. Calvin finished the bedroom first, and as he was completing the future sitting room, he saw that he would have a little lumber to spare. He extended the east wall by some three feet and then put two windows side by side to the east and smaller ones to the north and south. Although it was rectangular in shape, Nancy always referred to it proudly as her "bay window."

When it came to furnishing the rooms, the O'Neils were at a loss. They moved their bed and clothes chest to the bedroom. To take up the vacant space in the kitchen, Calvin built a cupboard. But for a long time, the sitting room remained empty.

One fall day was strange. Dawn had barely broken when a wisp of zephyrs arose from the earth, swirling the dust in Nancy's bare side yard. Gradually the breeze spiraled up to the trees, shaking first the yellowed leaves and finally the branches themselves. Calvin was toting two heavy buckets of foaming milk across the dusty yard and called to his wife.

"Come out, Nan. Git yerself a breath of this air. It's beginnin' to move."

Clad in her usual dress of dark calico, Nancy appeared in the doorway. The door with its fine metal latch was fastened back to the house with a crude rusting nail and hook. It had been too hot to keep the sturdy door closed, and Calvin had hung a piece of canvas from the old covered wagon over the aperture. It also now connected with the hook Nancy had found one day half buried in a pile of clods near where the well now stood.

"What's that smell, Cal?" she asked as she stepped into the yard, her usually busy hands bent backwards on her bony hips.

"Jest the wind, I reckon. Let me git to work on this milk so's you can have your cream."

"No, Cal, there is a different smell. Like melons. Or — it couldn't be rain, I guess?"

Suddenly, the wind reached turbulent proportions, maturing into a big bully wind. The branches shook down crowds of leaves which in turn shook themselves into deep eddies around the trees. A tremendous gust went straight for Nancy but decided to tease the canvas instead, tearing it from its hook.

Calvin had stepped inside in time to avoid having the buckets tipped, but Nancy lost her footing and sat down on the door stoop with a jolt.

"Nan," called her husband, "this here mush is burnin'!"

Mrs. O'Neil was an excellent cook and took great pride in her unusual ability to make the simplest food into a tasty dish. But already her right hip was sending sharp pains down her leg toward an ankle that was aching with a dull throb. Tears seemed to be standing in readiness in her eyes and as she heard that breakfast was ruining, they let loose in a cascade that covered her thin cheeks.

Nancy carefully lowered her neatly combed head onto her knees. An urgent blast of wind from the west blew the trees nearly to the ground, as if making a mocking bow before the weeping woman. The force of that wind came close to shocking the breath out of her, and before she could consider what she was doing, she stood up in a rage and stamped her foot hard upon the dusty stoop.

The resulting pain was formidable. The idea of broken bones flashed through Nancy's mind as she turned and hobbled into the house. A hatted head had been peeking around a fair-sized cottonwood tree and surreptitiously had witnessed the entire show.

Old Black Thunder viewed the farm as his own, even though not one of the improvements had come from him. Never had he offered to help the O'Neils with their labor. He had sat still while the couple had made repairs in the one-room house. He had stood, arms akimbo, while the young couple had devised a corral of sorts inside a crude dugout. He had absented himself completely when they, man and frail-looking wife, had worked the ground.

Nevertheless, he felt a certain attachment to his white friends. For friends they were, although the man was a more begrudging one than the woman. It had been he who had indicated to Black Thunder that the dugout could shelter him on rainy nights. No matter to him where he did sleep otherwise! Yet Calvin was never reluctant to share his board. Nancy always cooked enough for the three of them, and the red man seldom missed a meal.

On several occasions Black Thunder had definitely shown a sense of responsibility for these palefaces who had entered "his" house unbidden, much as he had entered the house when its white builder had departed permanently. They had taken over and he had acquiesced, waiting to experience the outcome. He quickly acquired a feeling of protection, almost of tenderness, for Nancy. As for Calvin, why should a man work so hard? But if he wanted to, let him do it alone. Yet when the water-witcher was missing his mark, why not help out? Shoot an arrow into the correct spot.

He was avid for Nancy's cooking. Her sweet dishes were enough to tie him to that family forever. No wonder when the braves apprehended the wagon and took her pies he made them return the sweets with the rest of the treasurable loot. How dear she was to him when she insisted that the Indians keep the delicious, juicy pies!

And how her sadness had touched his heart. She had called him "Friend," the same as she had called the departing newcomers. He was proud to escort her friends home safely past his tribe's hideout.

Had he not sat immobile for one long hour in front of her door one day to protect her from "man" while Calvin was away? Actually he did feel a certain attachment to his farm-mates. The tiny woman even tugged at his heart.

Though the wind was teasing the tiny woman Black Thunder loved the wind, actually loved it. He would stand with his body bared to it, letting it beat upon him. He would run with it, stand still and let it hit him head-on, lie down and tell it to blow across him. He would gather great gusts of it, sniffing to let it tell him where it had been.

On this day when the wind had announced itself in little zephyrs, the Indian had just awakened from his sleep on the north side of the cottonwood tree. He saw the milker hurrying indoors with his buckets and he saw the "cooker" standing on the stoop. As the wind developed, he grabbed for his hat and blanket. He let the blanket fall and stepped on it so that it would not blow away. He pulled the old gray hat down onto his head. It was a recent acquisition, and like a child he wore it constantly for a few days before losing it forever. He had punched holes in the hat, and through these holes he pulled his braids. Suddenly, he grabbed his blanket, gathered it to himself, and giving it a couple of fast twists, secured it under his arm until he could throw it into the dugout.

A mighty gust unsuccessfully tried to take both his hat and his blanket. A grin of enjoyment came to the face of Black Thunder and his spirit felt elated. He grunted approvingly when the wind tore the canvas from its mooring and laid it back upon the roof. There was one moment when the grin faded from his gorgeously strong face, the moment when Nancy was literally blown down by the wind. The moment was short lived. He saw that she was not hurt too badly, and then he saw her rise and stamp her foot in anger at the element that continued to persecute her.

Old Black Thunder could not repress his mirth any longer, but he knew it would shame his friend if he were to laugh outwardly at her defeat and defiance. As she disappeared into the house, he placed his blanket against the cottonwood tree to serve as a deadener of the sounds rising within him. His entire body shook as he stifled his merry laughter in the blanket. This helped, but did not satisfy the Indian's craving for a real laugh. Chivalrous manners did nothing but delay the act which he knew was Brother Wind's right. He sat down again beside the cottonwood and drew out some dried, stringy beef from a pouch that hung at his waist. This, he decided, would be his meal this morning.

Inside Nancy's kitchen, Calvin had eaten some of the unscorched part of the mush, and then he started up to the field. He had taken the canvas down from its blown position above the door, and he fastened it back as before. As the torn edge of the cloth fell upon the stoop, it covered Calvin's water bottle that he always carried with him to the field. After a few minutes of swallowing dry dust, he would be needing a drink of water from that bottle.

Black Thunder watched the farmer head west towards his work and he also watched the sky overhead. A storm was forming. The present stillness of the air bespoke a big storm. The Indian sniffed at the air. It was cooling — and cooling fast!

Inside the house Nancy rubbed her ankle with some lard and creosote, a mixture Calvin had prepared. The back rub he had given her seemed to have eased the pain in her hip, but her ankle still ached. As she was pulling on and lacing up her high-top shoes, she noticed Calvin's water bottle on the stoop.

"He'll need that," she told herself.

Black Thunder felt the wind and noticed the menacing cloud that had formed overhead. He stood to his feet and looked around in time to see Nancy starting across the side yard with the water bottle.

"No!" he yelled.

She stopped. With a series of utterances beyond her ken, he pointed towards the dark cloud and made motions as if to shoo the woman back to the house.

Nancy saw the cloud and fully realized that this was no time for her to be out of the house. But she wondered about her husband, wondered if he would take the cloud as a warning to hurry back to the house. In so thinking, she made a few quick steps forward as if to see Calvin in the field.

The watchdog Indian mistook her steps forward as meaning to brave the storm which he fully knew was upon them. He jumped in her path with a leap that was as graceful as an acrobat, and for a moment crouched as if to spring. Then he leapt to his feet, turned slightly forward so that his lean stomach was drawn in and his negligibly small hips arched back. With his hands he conveyed the urgency of going back.

Nancy watched as if entranced until a distant rumble was heard. Both she and the Indian stood in an absolutely immobile silence. The thunder rolled like a giant drum, becoming louder and more intense until the noise was directly over the farm. Then it let loose the loudest thunder yet, a roll and a clap. Loud, black thunder!

Nancy waited no longer. She turned as lightly as an aerialist and ran for the house. A few feet behind her ran her husband, already nearly soaked by the sudden short rain. Almost as soon as it had begun, that fall rain turned to hail. Meeting one another in the midst of the storm were Calvin and Black Thunder. They passed like the proverbial ships in the night, although much faster. Calvin reached the house and turned to see the Indian sliding around the entrance of the dugout.

Black Thunder stood bareheaded at the door of his sheltering place. He leaned out and caught some of the storm ice in his mouth. He saw his friend, the Wind, blowing the hail about. He remembered the joke that Brother Wind had played upon his friends and the canvas.

The Indian stood bareheaded in front of the dugout

and laughed. He let his laughter roll and roll until it healed him of all the pain he had felt as he restrained himself for the sake of his friend. The wind had given him a message: "Laugh, old friend. Be loud, Black Thunder!"

From then on the autumn days grew colder and Calvin's outdoor work dropped to a minimum. Not having used quite all of the materials he had gathered for the additional two rooms, the energetic fellow donned jacket, cap, and gloves one day and took several turns around the outside of the house, figuring how best to use the wood with the least waste. He had put in a solid wall to the north of the two new rooms, so that as yet the only entrance into the house was into the kitchen through the original door.

A short stretch of Indian summer days gave Calvin just the advantage he had wanted. To the east and to the west of the kitchen he built porches. He cut a hole in the east wall of the kitchen, and although the door he placed there was a crude one, it would serve as a way through the small porch to a lawn for Nancy.

For the long, narrow porch to the west he was able to board in the three walls up to waist high. He suspended slat panels from the roof which could be lifted and hooked to the ceiling in summertime. In this semi-room Nancy came to separate fresh milk, make soap, and do the washing. From the open walls of this west porch she watched her husband water the stock by morning and again at night.

From this workroom Nancy would see Black Thunder leave for the last time.

CHAPTER FOUR

## *The Raid*

*T*he year 1863 had seemed alternately to speed, then drag by. It was already more than seven months into 1864. On this day of August fourteenth, Black Thunder was not at the house for his customary evening meal. When Nancy dished up the fried rabbit and hot roasting ears, she filled his plate as usual and set it on the floor of the milk porch where he usually preferred to eat in hot weather. The food remained untouched and Nancy had a feeling that something was wrong as she set the plate back.

For over a year the old Indian and the O'Neils had been sharing the essentials of life with one another, not knowing for sure who was host or who was guest. Of course, the fourteen months of "proving and improving" the property had passed, and Calvin had a set of legal papers attaching the farm to his and Nancy's name; yet Black Thunder had been there before them. And perhaps it had come into his possession, too, by some previous, haphazard white-red trade. Nevertheless, they had lived a fairly compatible life together, each donating what he could to their common existence.

After Nancy finished the dishes, she thought about the ample supply of molasses they had on hand and she decided to make some doughnuts. The room was almost dark when she finished frying the last batch.

"It's a good thing I didn't make up more dough than I did. It's getting too dark to see," she told her husband.

# THE RAID

Getting to his feet and stretching, Calvin said, "Time to turn in, I reckon. Heap o' work in the field tomorrow." He supplied himself with a handful of doughnuts and turned towards the bedroom door. Before he could reach it, the canvas over the kitchen door moved and Black Thunder entered the room.

In the dim light he looked like a giant, a savage giant. From his face rose a mass of high feathers that encircled his head and fell almost to the floor. Instead of the old khaki shirt that had replaced the blue denim of the previous year, he wore eight or ten strands of heavy, bright beads, some round and black, some square and blue, and others the white teeth of prey. On each arm were circlets of animal hides, two made with beads and two with fur. A narrow loincloth covered his groin, and his legs were painted in angular designs of red and white. In his left hand was an arrowhead spear, and in his right hand was a tomahawk.

Placing his spear straight in front of his body, he opened his mouth to speak in the language that was so hard for him. With his tomahawk raised high over his head he said, "Friend. No field tomorrow. Stay house."

Before the meaning of these words could enter the minds of the stunned couple, the Indian had thrust his spear through several doughnuts. "Friends," he said once more and disappeared through the canvas door.

Nancy, now frightened, whispered, "What did he mean?" But Calvin only shook his head, a pensive look on his face.

"He said not to leave the house tomorrow," Nancy persisted.

"Oh, Nan," Calvin scolded. "Don't take that to heart. Couldn't you see that he was drunk?"

"Drunk?" Nancy repeated unbelievingly.

"War-dance drunk. Fergit it," her husband advised.

But Nancy did not forget it, and after Calvin was asleep she got up and tried to walk off her spell of nerves. On the milk porch was a bucket of water that had been pumped fresh the night before. Nancy slipped her cup into it to get a drink. Just as she raised the cup to her lips she thought she detected a scream. Her first thought was that it had come from a coyote, but upon hearing it again she sensed it was the yell of a savage man. As soon as she could get command of her frightened body, she ran to awaken Calvin. She insisted that he return with her to the porch. Together they listened, and from the north came a series of high screeches. Separately Calvin and Nancy envisioned the visit from Black Thunder a few hours earlier. Neither wanted to voice a suspicion.

"That's it?" he asked.

She whispered, "Yes, but before it came from over west."

An hour passed before the terrifying sounds could no longer be heard from either direction. Nancy still asked, "What is it?"

"Never heard the like," was the unsatisfactory response. But Nancy knew as well as her husband that those sounds were the cries of savage men preparing a raid. She did not close her eyes through that long night. Once more her heart sounded out the old question of why, oh why, was she left with no one in the world but this man who had brought her into this wild life to face a probable death by torture, and then lie there and sleep before its advent.

As soon as dawn had forced a tiny opening in the black curtain of night, Farmer O'Neil's blue eyes were open and he was ready to get to work. His wife pleaded with him not to leave the assured safety of their house, not even to milk the cows. But Calvin stubbornly prepared to do just as he did any day.

# THE RAID

"Then I'll go with you," Nancy cried. "What good would my life be after they had killed you?"

At this Calvin tried to appear disgusted with his wife and he laughed scornfully. But seeing her pitiful, weeping self pick up a milk pail, he tenderly took her hand as they made their way to the barn. With both of them milking, the task was soon done and they started back to the house. They stopped short in their tracks as an ear-splitting scream rent the air, echoes as it were of the previous night's fantasy. The sounds seemed to be coming from anywhere between a third to a half mile north of their house, and from a bit to the east. Then a little to the west. Then not so far east and then not so far west. Finally, a solid flank of yells exactly north.

Calvin O'Neil nudged his wife and motioned towards the house. But not until he pushed her could she move her feet along the path. When they stepped inside the house, Calvin motioned her to a chair and he took care of the milk and then prepared breakfast.

From inside the house, no sounds were heard and all seemed quiet. Occasionally, Calvin would walk to the door and listen. The shrill screams were still to be heard, but as morning progressed they were heard less and less often. Finally, it was noon and no sound had been heard for a couple of hours.

Calvin picked up his gun and told Nancy he would go out to take a look around. Nancy stood to her feet with a wild look of panic in her eyes. She put her hand to her face and it seemed to be burning. She wondered if it were red, and she pictured herself looking like Black Thunder in his war paint. She began to giggle, then laugh, then scream. When Calvin approached her, her tear-blurred eyes and her fright-twisted mind told her that he was an Indian. Her legs shook and her tiny body crumpled into a heap at his feet.

## THE FARM

Calvin lifted the little figure and carried her to the water bucket. It had tipped over and was empty. All he knew then to do was to lay her on the bed and slap her wrists in an effort to revive her. She was wearing a high-necked Mother Hubbard dress of dark calico. He unbuttoned the length of it and started to fan her. Seeing no improvement he quickly ran to the water bucket, grabbed it, and ran to the well. While he pumped some fresh water he detected a faint smell of smoke in the air, but he had no time to investigate it. Racing back to Nancy's side he doused cool water on her forehead and neck. She looked up into his face, gave a weak smile, and fell into a deep slumber. Calvin did not know that she had not slept in more than thirty hours, but he did know that hysteria could wear a person down to complete exhaustion.

Mopping his sweat-drenched face with a wet cloth, he stepped outside to cool off. The smell of smoke was definitely in the air, and it was rising from the west. He sensed that it was a prairie fire, the result, no doubt, of the insurrection that must have taken place the night before.

If the fire were headed their way, he knew he would need to protect his property, and even if it were not he should be helping others. But inside lay his wife, too ill to suspect the danger. Dared he leave her? As the odor became heavier, Calvin knew that he must rush to fight the fire.

Grabbing what tools he could carry he ran out across the barn lot and up into the field. As he ran he could see flames bursting up out of a not-too-distant field. His mind worked faster than he could command his body to move. Turning around he ran back to the end of the field that bordered the cow lot. In a few minutes the frenzied man had harnessed the oxen to the plow and had begun to plow deep furrows up and down the length of the field along the inside of the crude fence he had built. To the north and also to the south there were sufficient large rocks to stop a fire, but this stretch would lay the barn and even the house open to the blaze.

The greenness of the field was all that allowed him to complete his plowing before the fire reached close range. Exhausted, he pulled the team to safety and climbed up onto the wagon seat to watch the fire consume his crop. It was already a slow and dwindling fire straight ahead where the green corn stood, but to the north where the pasture was covered with prairie grass the fire burned hot and long. The plowed strips and the rocky ledge to the right kept the fire from invading the small patch of pastureland north of the barn lot and house.

Slowly, fatigued and broken-spirited, Calvin stepped to the ground and plodded up the shallow incline to the south. From that direction came the stench of burning flesh, and it came from the field that bordered the pigpen. A sort of moat encircled the pen and a rock fence lay between that and the field. Calvin could see that the pigs were gone. Walking through the sty he observed where a few rocks had slipped. Over them were the muddy marks of the pigs. They had stupidly run into the fire.

This meant a keen loss, both in food for the coming winter and in sales of ham and lard. But it was nothing in comparison with the loss of the crop. A year's work, getting the ground ready, cleared off and level. Plowing from sunup to sundown, planting the same. Fighting weeds and insects. It was his whole cash supply for another year. He and his land were defeated. Marred and charred by a mob of savages. What if they had protected his house? His fields meant more. Their yield could have built back another room to live in.

As he trudged slowly back to the area of the corral he envisioned what his field had been and what it had become overnight. He tried to picture himself ridding it of the burned stubble, and he wondered if he would have the heart to work it another year.

Once more he stretched out on the wagon seat to watch the dying fire. Occasionally he changed positions, but his eyes never turned from the desolate scene of the ruined crop and pasture land. At long length he fell asleep. Dreams came to him of the unbroken land covered with brush and stubble. He saw a distorted figure leading oxen over the rough ground. That picture seemed to last forever, a bent figure dwarfed by huge beasts. Then the dream changed to one of roaring flames which consumed first the oxen and then the pathetic figure. Calvin heard the man moan and groan, saw him twist and writhe, and awakened to find himself making the pained sounds.

His body was sore and felt as if it were baking with the heat. As he realized where he was he started to jump to his feet to survey the damage. But he was stopped by the pain in his back and legs. The ache was worse in his stiff arm. Aching, he slowly made himself rise to his feet. Seeing that the fire was only smoldering now, he left the wagon and turned toward the house.

For the first time since he had left the sleeping Nancy alone in the house, his thoughts included her. The memory of her, sick and frightened, came to him suddenly and forcefully, and he felt even sicker at the thought of his neglect. As quickly as he could move his parched and stiff legs, he strode to the house.

All was quiet. Fearfully Calvin tiptoed through the kitchen and on to the bedroom. There lay Nancy, much as he had left her. As he went closer to her he noticed that her usually colorless face was red with fever. Now and then she would flounce her arms about as if she were awakening. Each time, however, she would become still again in heavy sleep.

Seeing her in comparative safety, Calvin went weak

with relief. The soreness of his own body made him move toward the water bucket where he could get water to bathe his face and his arms. Also he knew that Nancy was apt to awaken and need water, so he picked up the pail and made an effort to get to the pump. But his strength was gone and his legs felt as if they were melting out from under him. As he set the bucket back on its stand, a wave of nausea engulfed him, and he stumbled outdoors. At first he seemed to be choking and his eyes lost their vision. Then for a few minutes after his sickness he felt better. He once more attempted the trip to the well. He needed a drink and this time he made it to the well. Soon after reentering the house he was again overcome by nausea. Hardly knowing that he was walking, he reached the bed and lay down beside his wife.

The movement of the bed awakened her and she started to rise. Her hands touched her dress buttons and she fell back onto the bed, trying to connect the present moment with the last one she had known. Then she heard Calvin's voice asking her if she was all right.

"Calvin," she said, as she took his hand, "what happened?"

"Indians scared you, I reckon. Do you remember blackin' out?"

"Did they come?" she asked breathlessly.

"Come to our field, all right. She's all gone now. Burned to the ground."

Nancy sat up on the edge of the bed and stared unbelievingly at her husband. He told her about the happenings of the previous afternoon and night. Her only response was a sad shaking of her head and a bare whisper, "All our work."

Slowly she got to her feet and walked out to the kitchen. Her head was aching violently and she needed coffee. She wished for some mixture that contained real coffee, but all she had was coffee essence. She decided to make a strong brew of it and she set it on the stove to boil. But the fire was dead and no wood lay near the stove. She felt much too weak to try to go outside for some kindling. She would have to ask Calvin to do it.

Turning towards the north her eyes and mind took in the sight of the two new rooms. There was her precious sitting room, empty as it was. And there was her bedroom. She walked to the doorway and leaned against it, smiling at Calvin as she said, "They didn't burn the house anyway."

She looked a little bit triumphant and even somewhat happy, feeling that what mattered most had been spared. But Calvin wore a look of bitter grimness, realizing that defeat had come to him and to his land. All that was left was what they actually could have done without. Their habitual visages had been reversed as each one's mind had taken hold of the facts in his own way.

Suddenly there was a sound of voices outside the house. It was the Katchalls, and as Julius entered the kitchen he called back to his wife, "They here."

Frau Katchall walked into the room as fast as her weight would allow. She put her arms around Nancy and said, "Gott in Himmel. You are safe." The children looked on with big grins as they heard their father tell the O'Neils that nothing would do the woman until they came over.

Henrietta could see that her friend was weak and that no food had been prepared. Capably she sent the boys for wood, and soon she had the coffee boiling over a hot fire. While she cooked a pot of hot mush, her husband told the news. They had heard some of the noise that the Indians had made in their raid the previous morning. Later they learned that there were several rampaging tribes, each waging war in a different locale and finally converging in one big assembly in the valley.

# THE RAID

Julius had taken his gun and stealthily made his way over to the main road leading from Marysville. There he encountered state troops that were heading back north. He learned that Hod Beach had ridden into Marysville to give warning of the coming onslaught, having learned of the plans from his squaw and wanting to gain revenge for repeated thefts. Then he had made his escape, leaving Kansas as a reminder of his past. In the meantime soldiers marched into the valley just as the Indians' final merge was accomplished. Before the marauders could attack Pierson's mill, the soldiers had killed one of the chiefs, Black Thunder, and countless braves. The ones escaping vanished into the hills northwest of the valley.

What facts Julius could not supply, the O'Neils learned later. Royal Lee, the water witch, had been scalped and his body thrown into a well. The pump crew for which he worked ended up attached to their own machinery, tortured to death. It was Ed Winchell's cabin that they had set afire, which had ignited O'Neil's field. As for Hod Beach, his old shack had been burned and his squaw dragged into the fray from a rope attached to one of the brave's horses.

When Julius mentioned Black Thunder's death, Nancy gasped. She remembered the warning he had given them, and she knew that her house had been spared by his decision. He was their friend. But Calvin experienced no such feelings. He remembered the labor he had put in on his crop, and he knew that his land had suffered because of the old warrior's acts. He was an enemy well lost.

All afternoon the two families talked of the terrors of the red man. Finally, the Katchalls got into their wagon and rode away, knowing they must arrive at home before dark. This time there was no protecting figure to shadow them, or on the other hand, no menace to be protected from.

## THE FARM

The O'Neils turned towards their field and walked over to survey the damage. All was ruined. As far as they could see was charred earth with a few stubbles here and there. No pigs. No corn. No pasture grass. Backbreaking work for nothing. All to do over. Tears came, first to the woman and then to the man. Life could be so hard on helpless humanity.

This was the end of Indian terror as far as the O'Neils directly were concerned. But the little settlement in the valley was not entirely free from the onslaught of the red man's anger.

CHAPTER FIVE

## *Braving the Cruel Forces of Nature*

During the winter of 1867 the Central Branch of the Union Pacific Railroad had finished laying the rails from Atchison to Pierson's mill. It was quite a thrill for the settlers to see the train puffing along the wooden tracks, because embosomed within were possible letters from home, rare packages carrying precious cargo, and perhaps even an occasional visitor from back East.

The thrill was tempered with chill, however, as yet more Indian raids were inflicted upon the valley — raids upon the railroad itself. For the Indian, the snorting machine created no soothing pictures. Instead he saw it as a weapon from the evil spirits themselves, sent to torment and possibly to eradicate the red men. At first the braves would lie concealed behind rocks to watch the ogre chug by. But it was to them a major cause for alarm, and soon the problem of what to do about it became the bond that linked together the various tribes in eastern Kansas.

Certain tribes made plans to strike at it physically, tearing it up. Others truly saw in it the overpowering might of the white man and sought to teach him a lesson by hindering his progress. But they were united in their idea of how to accomplish their varied aims. Before the dry heat of summer spread its parching fingers over the Kansas plains, several tribes had rushed upon the menace and had done whatever damage they saw to do.

In the valley below the O'Neil farm, scores of braves

THE FARM

had torn up tracks; others hammered and beat on the train and its engine; still others attacked the crew and killed whomever was in sight. Every shack or house bordering the railroad property was set on fire, trapping many a family too frightened to leave its walls and face the fright-maddened red men.

That attack on the railroad and its workers was the worst massacre ever visited upon the settlers of the valley, and it was the last. State troops drove the Indians to more unsettled territory. However, it took years for the settlers to realize that they were free from the savage terror. They had to live as if at any moment the tartars could appear and encircle them with their painted bodies, emit their bloodcurdling yells, and brandish their killing torches.

For a long time, Nancy O'Neil lived in dreadful fear of the red man. Calvin never left her alone but that her ears began to listen for a scream or a stealthy footstep. She never accompanied him to the field without keeping her eyes alert for a figure behind a rock or for a shadow. She never returned to her house without fearing what might have taken possession in her absence.

But the O'Neils were blessedly excused from this worst of all terrors, that of receiving harm and damage from another human being. However, there were other hardships visited upon them which at times seemed beyond their strength to bear.

When the O'Neils had been living on their farm for a little over four years, disaster struck in the form of grasshoppers. The soaring creatures appeared from the north as a dark cloud of floating seeds. Part of the cloud disintegrated and took to the ground, covering it from one to three inches deep. Before the day was over, the harvest was destroyed. Once more Calvin and Nancy, along with many other settlers, faced a winter with no grain in the barn, and no money.

Although winters were severe, the one which followed the grasshopper blight was cruelly fierce. Snow began falling early in December, covering the ground with its thick blanket. The freezing wind drove the flying snow until some of its banks were as high as the house they encompassed. Two and three days of every week brought snow and wind to the Kansas homesteads.

The year 1868 was blown into Kansas by an ice-dust-wind storm. Nancy and Calvin even felt the extreme cold within their three-room house. Water froze in the bucket, as did the potatoes and pumpkins which they were using sparingly against worse days ahead. The young couple moved into the kitchen since it lay to the south, and put their bed close to the fireplace.

Calvin thought of the cows and made an effort to reach them. But before he could reach the barn lot, his face was covered with ice. Thin ice pelted his clothes and almost instantly transformed them into icy armor. He turned around to try to protect his eyes from the gale and snowstorm, and still see his way home. By the time he reached the door his eyes were frozen shut.

When the blizzard had subsided, Calvin again started to the barn. It was impossible to hurry over the slick ice, and part of the way he could only progress by crawling. Upon reaching the sod structure he found his team of oxen about to suffocate. The snow particles had been fine enough to penetrate every crevice. Calvin used his shovel to knock loose enough ice to let the air penetrate.

Then he sought the dugout stable where the cows were sheltered. There the sight that met him was almost too much for a man already sick with the cold and the feeling of defeat. One cow and her calf lay frozen to death. The other one had evidently stayed longer within the shelter, as she still breathed. But her body was thickly frosted in ice, and her head was covered with such a mass of ice that she could not support the weight. Calvin used his shovel to beat off the case of ice. Then he pulled the numbed cow into the dugout and rubbed her with straw. Several inches of snow had blown into the cave, but still it was shelter for the pained animal.

Calvin could not battle with dead animals alone, and he had to go to the house for Nancy. She felt that she was freezing to death as they pulled the beasts into the barn and prepared them for meat. Food was too scarce to waste a frozen cow.

When the December storms had begun, Nancy had taken her few hens into the milk porch for warmth. Still they could not survive the tremendous blizzard and for the second time she lost her entire brood of chickens.

It took days for Calvin to work his way up to the pigpen. It took several more days for him to locate any of the litter. Two of them had found a deep drift and had burrowed under it for shelter. They were hungry, but alive and fairly warm. The others were never seen again.

Nancy and Calvin sat day after day in dread despair. They wondered what was to be the outcome. What was to be gained when such back-breaking work as they had known could be so quickly destroyed, first by fire, then by grasshoppers, and latest by blizzard.

Why, oh why, thought Nancy, had it been her lot to lose her parents and end up in such a place of desolation? Even her frame house was no consolation because the cold had frozen fast the doors to the north rooms.

Calvin sat as though whipped. Leaning forward with his elbows on his knees, he would sit for long spells at a time, occasionally running his rough hands through his black curls. The twinkle in his round blue eyes was gone, and no cheery whistle came from his lips.

"One thing to be glad for," his wife was thinking, "no children to suffer in this place of cold and want." And

Calvin was thinking, "Right now we need a couple of boys to give us reason for pushing on."

But with the passing of time, despair often lifts. And with no tangible cause for elation, hope can spring up in a man's being to urge him on. As soft breezes began to blow across the plains, Calvin and Nancy once more set their hearts on the season's harvest, and their energy to the task of making it a good one.

This particular year, through some odd hunch he felt, Calvin put in less corn than usual. Feeling somehow that poultry was safer, he swapped work for hen eggs with first one setter and then another until he had a large amount of poultry. What had made him do such a thing was a riddle to Nancy, but on August the third they were glad for his actions.

It was a bright and sunny day and Calvin had just finished his noon-day meal. He started to the well for some water when he noticed that the sun was darkened by a cloud of dust. He watched the cloud come slowly closer, when suddenly behind him he heard Nancy cry, "Grasshoppers!" He turned and saw her heading for the chickens.

"Head 'em in," he called to her, running swiftly to the other side of the house to drive them back. Together they chased their chickens into the crude soddie Calvin had built for their shelter. Before the task was completed, the insects had descended. As the hoppers hit the house, the noise was like hail. The air was thick with them.

Calvin ran across the lot and up to the pigs, but he saw that they were gorging on the grasshoppers as fast as they could eat. Before he could get back to the house, his pants and shirt were nearly eaten off him.

The insects stayed for three days, eating every green thing, even foliage off trees. They ravaged the garden, eating turnips and onions right into the ground. When they fed on the corn, the noise was as great as if a herd of cattle were eating. Even the weeds were destroyed.

When the hoppers departed, they left only ruin and desolation. Calvin had lost his entire crop, small though it was. There was no food now for the poultry, so he and Nancy prepared them for market. It was their only resource to cling to. Some of the pigs had gorged until they burst, and all of them had eaten until their meat tasted like grasshoppers. They were unfit for market, but the O'Neils had no recourse but to eat them. Calvin and Nancy lived on pork that entire year.

With money from the chickens, Calvin decided to buy seed for a wheat crop. Not knowing what freak of nature would strike next, they planted the wheat only to see it blown away in a dust storm.

Wearily Calvin went along the sides of the road shoveling up the grain deposited there by the storm. He salvaged enough to plant a small crop the next year. For seven years he and Nancy had seen failure and destruction in almost all that they had undertaken. But there was still the eighth year to hope against.

## CHAPTER SIX

### Childbirth

At the O'Neil farm and in the little village in the valley, all years were not dark. And those that were not shone out in bright contrast. The brightest of these was Christmas of 1869.

Almost a year prior, the townsite of Waterville had been surveyed by the chief engineer of the Union Pacific Central branch. The village was to be laid out in the valley which nestled under the hill just north of the O'Neil farm. Almost immediately business houses and dwellings were built. Among the first buildings to be erected was a frame schoolhouse, built at a cost of one thousand five hundred dollars. In this building was to be held a community Christmas Eve celebration.

Nancy was looking forward to the occasion with the first real enthusiasm she had shown in years. During the fall, Calvin had worked in the village helping to build this one's home and that one's store. His wife had disapproved of these daytime treks of Calvin's to the point of being sulky upon his return at night. During his absence she would continually brood about the emptiness of her sitting room, and fret about the work to done at the farm.

And indeed, Calvin himself became fussy over the work which took him away from home and his land. He would often return at night in such a state of moodiness that he was worse than no company at all.

Nancy knew in her heart that he was laboring for items that were badly needed on the farm. And yet she held the grudge for it against him until he brought home the news of the Christmas party. With the news he handed her a parcel, the first present in such a long time that she hardly knew how to take it. Tears came to her eyes as she fingered the gift, a bolt of brown wool for a new dress.

Like a child she ran to the bedroom where the chest stood that they had brought from Missouri. There in the bottom drawer wrapped in a piece of blue muslin lay the lace collar she had worn on her wedding day. Running back to the kitchen with it in her hand she said, "Can't you see how rich this lace will look on the new brown wool?" She went to Calvin and gave him the first good kiss he had had for a while.

"Well, I'll be a bat," Calvin said as he ran his right hand through his thick, black curls. "If I'd a knowed this, I'd a wove you some wool."

Nancy tenderly laid her gift away until she could transform it into a dress, and then continued preparing her meal. Speaking as much to herself as to her husband she said, "And you've been working so hard to buy me a dress. Didn't you get anything for yourself?"

"Yup," came the reply.

Not really hearing his reply, she said, "You ought."

"I did," he said with emphasis.

"What?" she asked unbelievingly.

"I did," he repeated.

"Well! What?" she demanded.

With a big grin Calvin walked past his wife and started out the door. Nancy left the stove and followed him out. As he passed through the milk porch he stopped to pick up the water pail. His wife almost stayed in the house, leery of any new surprise outside. But the grin he gave her as he turned toward the well beckoned her on. The sight she saw brought a delighted gasp to her throat.

"Oh, Calvin, is it ours?"

# CHILDBIRTH

"All ours," he told her. "The finest team of horses I could get fer usin' my two hands to build."

But what took the woman's fancy was not so much the team of horses as the little black buggy behind them. It was a cold December day, but for once Nancy did not notice the cold as she climbed into the buggy to be driven around the barn lot.

She was still happy and warmly aglow as they drove down to the schoolhouse a few days later. Calvin drove the buggy out of the way in order to show his wife some of the work he had done in bartering for new items. The post office had been the first job he had helped on. Then had come two hotels, the Eagle House and the Lick House. On the street leading to the schoolhouse was a frame house he had helped erect for a family named Foerster. The site was getting to be quite a village, and Nancy was proud of the part her husband had in its start.

The party was a success. It consisted of a spelling bee, pumpkin pie and coffee, and lots of visiting. The Katchalls were there with their family, now consisting of seven, the two new additions being a set of four-year-old twins. Clara, a big girl of ten years, spent most of her time looking after Kurtz and Katrine. They were a lively pair, and Nancy noticed that Calvin could hardly take his eyes off them. She also felt that she had mentally tagged each child there with its proper family and had found themselves to be the only childless couple. She quickly summoned to her mind the actuality of the hard life they were forced to live, and she was thankful not to have a child to worry over.

The evening ended with an announcement from a Mr. Frank Root that the first regular edition of the Waterville paper (later to be called the *Telegraph* by everyone but Calvin who referred to it as the *Astonisher*) would be published and ready for sale on New Year's Day. On the way home, Calvin drove the team by Root's one-room newspaper office and home.

"I fitted them doors and windows fer him," he told Nancy loudly.

Getting only a nod as a response, he went on, "Wasn't worth much." Still no answer. "Oh well," he concluded. "I guess it won't be so bad readin' the paper regular."

"Oh, Calvin," laughed Nancy, slipping her hand into that of the master barterer.

The early months of 1870 saw Nancy O'Neil's improved state of mind continuing. She was happier than Calvin had ever seen her. As spring progressed she took heart in the work of the farm. Her husband had continued his labor-for-goods swapping and had earned a fund of supplies on which he could draw as wants arose. This security gave her a new willingness to team her energy with that of the soil, and until the middle of June she worked harder than before.

As usual she helped with the plowing and cultivating, made easier this year by the team of horses. But this heavy task was only her sideline. Her main interest lay in the garden which she tended alone. She raised onions, cabbage, squash, and beets. Everything that would grow was planted in her garden except potatoes. These were in Calvin's garden, and his crop grew alongside Nancy's. In the last hour before darkness the curly-haired farmer could be seen working amid the rows of potatoes, and directly to the north of him would be his frail-looking wife, using strength equal to his to weed her garden.

Warm spring seemed to change overnight to hot summer, quite similar to the first year they had lived in Kansas. As heat descended onto the O'Neil farm it stifled every bit of cooling breeze, and Nancy felt ill. It overtook her one morning as she started to help with the milking.

# CHILDBIRTH

Calvin had traded the oxen for a Jersey heifer and a stretch of carpentry for its twin. They were beginning to give as much yield as the fine old milk cow that had survived the blizzard.

To tend to the milking had been a task which Nancy never minded. But on a certain June morning the heat turned the scent of the rich milk into an odor unendurable to the small woman. Calvin saw her lean her head against the side of the cow she was milking and all but sway to the ground. As he helped her to the house he said, "You'd better eat a bite before you try to work in this heat."

She nodded, but as she entered the kitchen the smell of the coffee she had left boiling finished the job of complete nausea. Before it left her, she felt as if her entire body had been torn inside out. Slightly relieved, she began preparation of the huge breakfast they were used to eating before going to the field. She managed to eat some and even started out with her husband. But the nausea returned and she was forced to go back.

As the days passed Mrs. O'Neil grew gradually worse, hardly able at times to raise her head from the pillow, and not caring if she did or not. The work was left for Calvin and not even the thought of her garden could get Nancy to her feet. Morning would slowly work its way into afternoon, and slowly into evening. Dark descended so slowly that it seemed to be forever on its way. But when at last it enveloped the earth within its blackness, the sick woman was able to bury her head in the pillow and find relief in sleep. Then all too soon the tortuous daylight would force Nancy back to long hours of suffering.

So this was what it was like to be pregnant! The thirty-year-old woman had never known illness, and to be besieged by this violent sickness from morning to night, day after day, was wretchedness almost past endurance. Her husband was sorry for her in her misery, but there was so much work for him to do that he saw her only a brief part of each day. At night he would reach her side in utter exhaustion, and he would fall asleep immediately.

By the Fourth of July she had been suffering for over a month, and there was no indication of any letup. Earlier in the year she had made arrangements to meet Henrietta Katchall at the Independence Day celebration to be held in Waterville. But on that bright and sunny day she could hardly turn in her bed without vomiting a green bile-like fluid that seemed to tear out the lining of her stomach before leaving it.

When she heard her husband separating the cream from the milk, she called to him and reminded him that it was the Fourth of July. He muttered something and she told him that he must go into the valley and tell the Katchalls of her sickness.

"Nan," he said wearily, "the wheat cannot be left. I must work today."

The woman begged him to go, and finally broke into sobs which brought back the nausea. But the farmer had worked so hard, and now that the crop was at the point of harvest he could not possibly leave it. Working by himself as he was, every hour counted. Any delay might mean complete ruin by any of the blows of nature, from heavy rains to grasshoppers. He could not allow himself to yield. He could only turn his back on his stricken wife and hasten to the ripe field.

For hours she lay in misery, her face wet with tears. She felt so alone, uncared for. All the bitterness of her past griefs came to her mind, and she began to relive the horrors she had endured since her marriage. Lying there, living as it were through the freezing cold of blizzards and the burning heat of the windstorms, through the frightening spectacle of the Indians and the desolation of the insect-riddled farm, she failed to hear the steps of her

# CHILDBIRTH

friend as she entered the room.

For the second time in the life of their friendship, Henrietta had sensed the need to go to her "freund," feeling that there was something wrong. Oh, the joy that Nancy knew when she realized that Henrietta was there! A being who was interested in her, a woman who could sympathize, a friend. Her tears increased as Henrietta sat at her side and rubbed Nancy's aching forehead. The realization that someone cared was as hard to bear as the wounds of seeming neglect.

While Henrietta soothed Nancy, Julius and the boys found Calvin in the field and lent their hands to the work of the harvest. As Henrietta stood up for Calvin's cause, making Nancy see that the work had to be done, so Julius let Calvin know that a woman in Nancy's condition sometimes needed a certain amount of coddling. When the Katchalls left for their own home, the situation was greatly eased at the O'Neil farm.

As summer turned to fall, Nancy's nausea gave way to extreme discomfort. Each time she stood, a weight pressed down on her small frame and cutting pain would cross her body. One Sunday in October Calvin rode horseback to Katchall's soddie and explained the situation to Henrietta. In her German way she listened to the account in silence, occasionally nodding and whispering, "Ja, ach ja." After Calvin had finished his descriptions of Nancy's pains, Frau Katchall handed him a mixture of old herbs to brew, and in broken English prescribed hot applications.

Calvin rode home as fast as the horse could make the five-mile trip, and then administered the suggested treatment. The combination of the hot water packs and the steaming teas had a marvelous effect upon the patient. Before long, Nancy O'Neil was able to walk around and do some housework without too much pain.

But with each season there came a new affliction. As winter began to poke its icy fingers into every small crack and crevice, Nancy took down in the back. At times she felt as if the small part of her back were pulling away from the rest of her body. The weight of the child bore down so heavily that the mother's small frame could hardly support it. Nancy was forced to lie down a part of each hour, a pillow pressed against the small part of her back.

During the winter season, Calvin's work lessened and he had more time to sit inside with his wife. Sometimes he could hardly wait until he would have a boy to romp with, to have as a constant companion. He recalled his own youth and how he had been forced into work that he dreaded. It came to him to wonder if his son would ever dread farm work. Surely not, not this boy. He would be a farmer from the start. Watching his wife go through so much agony he had a suspicion that this would be their only child. The thought saddened him. He seldom had been able to take a walk out over the land without visualizing what it would be with four or five boys rollicking over it. One might have taken more to the stock, another to the seed-bearing sod. But all would have sent their roots down into the ground to become part of this farm, or another one.

The second week in February found Nancy so uncomfortable that she had taken to her bed, vowing not to leave it until the birth of the child. Then she began to say that she would never leave it alive, and Calvin hurriedly rode northward to the Katchalls' place. Henrietta told him to get on home and make Nancy get to her feet and stay there. She promised to follow him early the next morning.

When Henrietta Katchall arrived at the O'Neil homestead on the eleventh of February, 1871, she found Calvin pacing the floor and Nancy still in bed, still vowing that she would never move. As soon as she had sent her family on its way back home, the improvised midwife took things in her own capable hands. She bade Calvin keep the fireplace roaring and a kettle of water ever boiling. Then she lifted the ailing woman out of the bed, wrapped her in a warm blanket, and set her to walking. All day she kept her on her feet, or sitting in front of the fire. By nighttime Nancy was in a sound sleep as soon as Henrietta allowed her to lie down.

At four o'clock Nancy awakened Henrietta who was lying beside her. "Henrietta!" she screamed in alarm. "I don't feel right. Something is wrong."

Her friend and doctor once more got the patient to her feet. "Es ist hier," she announced at the conclusion of the examination, meaning that the long fight had begun.

While Henrietta fed Nancy hot tea and walked with her up and down the kitchen, Calvin sat in front of the fire and shook as if he were out in a cold blast. Occasionally, a cramp would hit Nancy. She would maneuver a sort of smirk towards Henrietta and say, "This is nothing." At this the man would smile in relief, while the more experienced helper would shake her head and grimly mutter, "Ach, du lieber."

At noon Nancy was sitting to eat a bowl of hot mush when she felt a pain much stronger than she had previously known. It sickened her. She pushed the bowl of mush back and started for her bed. But Henrietta was of the walking school, so she quickly grabbed the arm of her patient. Together, they plodded around the kitchen and into the bedroom and back.

Calvin had blazing fires in all three of the fireplaces, but the sitting room was still bare of any furniture and Nancy had no joy entering it. Cold air was sifting in at the bay window anyway, so Henrietta bade Calvin to shut the sitting room door and snuff out the fire. He did shut the door but he let the fire burn on. The room was somewhat drafty but it served as a haven to him where he could lie down before the fire and be out of sight of his sick wife. Poor Nan, he thought. This one would be enough. He never wanted to see her in such torture again.

Calvin was roused by hearing his wife give a pitiful cry. He dashed into their bedroom to find her lying on their bed and clutching her back.

"Nichts rub," Henrietta warned Calvin as she tried to get hold of Nancy's little hands.

On seeing her husband, Nancy thought to ally him to her side: "Cal, my back, my back. Please let me rub it."

Calvin turned to Henrietta with a sick look full of pity and asked, "Can't she?"

"Nein. Nichts rub," was the stubborn reply.

For over an hour Nancy writhed in pain, keeping up a constant whine that her back was breaking. About two o'clock the pains began to cut across her stomach, and in their violence the back was forgotten. Calvin looked ill and Henrietta ordered him to see about the fires. Nancy's whines increased to sharp cries and then to shrill screams.

For two more hours the three of them went through the torment, one being physically torn, the other two having their nerves frayed. Soon after four o'clock in the afternoon Henrietta called for Calvin to bring the hot tea she had been brewing. Together they had forced little sips down the throat of the woman who was tearing at the bedclothes and making frantic efforts to loose herself from all earthly ties.

The brew began to have a soothing effect, and as Nancy relaxed, Henrietta pressed on her stomach. The ache from that pressure was terrific. Nancy summoned all of her strength to her arm as she slapped the face of the one who had caused the pain. Henrietta did not flinch but ordered Calvin to hold Nancy as she poured more brew

## CHILDBIRTH

down the desperate woman's throat. As once more she relaxed a moment, Henrietta gave another strong punch on her stomach. Nancy felt her body open up to give way to a force that seemed in her semi-delirium to be her large ruby red bowl. The relief that came with it was enormous, and Nancy closed her eyes. She seemed to have faded away for all at once she felt another strong push, and the last part of the birth was over.

Before she fell completely asleep she heard her husband say in a sweet and gentle voice, "We have a baby girl, Nan."

## CHAPTER SEVEN

### Tinette O'Neil

The seventh day of May, 1873, found the O'Neils preparing to entertain in honor of their tenth anniversary. At the age of forty-one, Calvin looked much the same as he had on his wedding day. Except for the occasional bouts of soreness in the old arm wound, he really seemed more sprightly than ever. His back was as straight as a papoose board and he walked with a firm, quick step. Although his face had become like tanned leather, it was mostly unlined, and from it his eyes shone out as large and bright blue as ever. So far no gray had appeared among the tight black curls. He wore a blue checkered shirt which was open at the throat, and his homemade grainsack trousers were partly covered by the high leather boots he wore. His hands looked older than himself. They were rough and hard, showing the difficult times they had been forced to work through, and the thumb joints stuck out sharply from the hands.

Calvin was helping Nancy extend table boards to accommodate their guests. Nancy O'Neil looked older than her thirty-two years. Her entire appearance was that of an older woman. Her brown hair was liberally filled with gray. Her small eyes were more deep set than ever and were marked by crinkles and lines. Her posture was poor and indicative of work too heavy for her tiny frame. Her stomach protruded a little, but in spite of it all she did look neat in a beruffled dress of green calico.

As the couple arranged the tables in the sitting room

where the guests were to sit, little Tinette sat in a small rocker near the bay window. The two-year-old child had been named Tinette in honor of the baby that Henrietta Katchall had buried along the old pioneer trail. Tinette was a happy little girl who offered no trouble of any kind to her hard-working parents. At the same time she followed all of their work and movements with intelligent curiosity.

The guests began to arrive, each family bringing its own dishes and flatware, and some bearing gifts such as turkeys and lambs. The Katchalls, all seven of them, were there, scrubbed until they shone and happy until they glowed. They brought big wooden trays of Kaffe Kuchen.

The finest gift of all was from Sam Stanton, the boy who lived across the road. A month previous he had staked a claim on the land that stretched along the east side of the trail that bordered the O'Neil farm. He was only a lad of some twenty years, and he had with him his three orphaned brothers. Two were just younger than himself, while little Jim was Tinette's age. Nancy had practically adopted the four boys and she helped them in every way she could. The older boys in return did a great deal of work for Calvin.

The gift Sam Stanton presented to Nancy was a gold brooch. It was a beautiful pin, set with four small pearls which encircled an emerald. He said that when his mother was giving birth to Jim, a stagecoach had stopped in their little village. When the travelers had resumed their seats and the wagon rolled on, there lay the brooch beside one of the wheel ruts. His mother, weak from her ordeal, had motioned for him to place it in a box. It had lain there until tonight. Nancy could not hold back a few tears as she fastened the pin onto her dress. Impetuously she gave Sam a tight hug. This brought happy laughter to the group, and preparations for the supper were resumed.

# THE FARM

Calvin carried Tinette around until it was time to place her with the other children at the low table he had made for them. The children were given their plates first, and then the grown-ups sat down to eat. Nancy was too overjoyed to eat much, but Calvin only stopped at intervals when it was necessary to throw back his head in a hearty laugh. Tinette, to the great amusement of the older children, found it delightful to echo each of his laughs with one of her own.

Good times prevailed. Things had begun to look up for the O'Neils. The Indians had not been hostile around Waterville for the past five or six years. No more prairie fires had occurred. The insects were appearing less often and the continuing growth of the trees was decreasing the force of the windstorms. True, the winters were long and cold, but Calvin was ready to build a new barn for the stock, and he decided to put shutters on the house windows for extra warmth. Despair had almost driven them out in former years, but the O'Neils felt the worst was over. Life stretched out ahead. A good life, with little Tinette to live for.

Calvin and Nancy O'Neil had been among the first patients who sought the medical advice of Dr. Hugo Pattison. He had arrived in Waterville in 1871, already past middle age. Living with him was his daughter, Mary, a young lady still in her teens, and it was she who opened the office door for the O'Neils. She struck Nancy as being haughty because the girl merely turned and silently left the room unsmiling, not even stopping to look at the child Calvin was carrying in his arms. Mary's behavior made Nancy wonder if the doctor would be as unfriendly as his daughter. Actually it turned out that neither was and before long all were good, if not close, friends.

Mary had been expensively schooled in St. Paul,

Minnesota. Being shy of nature she had not become a part of the social life of that city and had known little popularity among her school sisters. Upon finishing the prescribed course of study required by the Academy, she returned with great relief to live with her widowed father.

As time went on, the good doctor became very fond of little Tinette O'Neil, and when his daughter expressed a desire to tutor the child, he broached the subject to the parents. As a result Tinette studied with Miss Mary for several years. There was not much affection between the two, but Mary did take joy in her pupil's keen aptitude. Finally, Tinette had entered and was graduated from Waterville High as one of its brightest students.

On the night of February 12, 1891, the Pattisons gave a party in honor of Tinette's twentieth birthday. Tinette was a pretty girl. She had her mother's small nose but the general shape of her face was that of her father — the high cheek bones, perfectly bowed though rather thin lips, and a high, smooth forehead. Her complexion was white and clear, with just a few freckles from not being too careful about wearing a sunbonnet. She wore her brownish-auburn hair in braids that were wound about her head like a crown over her large, deep blue eyes.

"Having a good time, Honey?" Dr. Pattison asked her. "And what did you think of our new dentist?"

"Dentist?" Tinette whispered with a bewildered pucker.

"The young man I made you acquainted with, Earl Norris. How'd you like him?"

"Oh, Dr. Norris. He seems all right."

"He is all right, Lassie. He can read a book the cleanest of any man you ever met. Mary says there's nothing in history he can't recount."

"He's all right," Tinette agreed as she turned enough to get him within the scope of her gaze. "From here it is surprising how much his face looks like that of Thomas Jefferson. Does he ever wear a wig?"

They parted laughing and circulated among the guests.

Earl Norris was a young man who loved beauty, and he could not take his eyes off Tinette. Her gay laughter and quick manner of speech set his heart to beating excitedly, and her large eyes held him fascinated. Her excellent figure made his blood tingle and he wanted to touch her. He tried to conceal his interest, but when he learned that she was to spend the rest of the week with the Pattisons he could not help but betray his feelings.

"This is better than I'd hoped for, Doc," he whispered to his host.

"She doesn't even know that you're alive, Boy," was the older man's amused answer.

"Maybe I can show her we're both alive," the dentist warned, eyes a-twinkle.

"Go to it. Just take care. That's all I have to say," responded Dr. Pattison.

When the guests had departed, Tinette was a little surprised to find that Earl Norris had stayed on. He was standing in the parlor by the fireplace, one elbow carelessly resting on the mantle. His frank smile rather disarmed Tinette, and she paused in the doorway not certain what to do next.

"Come on in, Honey," the doctor called out from his big rocker. "Tell me if you're glad to be twenty."

The young man at the fireplace softly murmured, "Twenty."

"Yes, twenty. I didn't see her the day she gave her first squall, but gosh-soon after. Like your party, Lassie?"

Dr. Pattison's question brought Tinette back to herself and as she crossed the room to go to his side she exclaimed, "Oh, Doc! It's the happiest day of my life. The party, all those presents! You and Miss Mary are too good to me." She turned to smile at Mary Pattison, but Mary was totally absorbed, staring at Earl Norris. Tinette turned back to Dr. Pattison and went on. "All you have done for me these twenty years. It's enough to bring a tear to my eye."

"Now don't go messing up that pretty face with unqualified tears," the old doctor said affectionately.

"Unqualified?" she asked.

"Absolutely unfit for the occasion."

"Oh Doc, you are a peach," laughed Tinette. "Instead of crying I'll see if I can begin to straighten things up." As she left the room she glanced at Mary and received an approving nod, but no smile. Mary again turned to look at Earl Norris who, she decided, was about her own age. Maybe younger. She wasn't sure. Her father gave a sideward nod of the head as he, too, looked at Earl Norris. The young man strode out of the room and found Tinette in a room across the hall.

"May I help you move those chairs around?" Earl offered. Then after an awkward pause from the silent girl, he added, "I'd like to."

Young Dr. Norris was very appealing. He was neat and attractive, his etiquette beyond criticism, his manner gracious. It would have been nearly impossible not to respond to him in good humor.

"Thank you," Tinette said. "Perhaps we can get this done before Miss Mary gets around to it."

Tinette and Earl straightened the room in complete silence. In a few minutes Dr. Pattison joined them.

"Mary's headache drove her upstairs, and I think I'd best be getting to bed, too," he said, looking at his pocket watch. Earl thought it was meant as a hint.

"Let me tell you good night and I'll be heading for home and bed myself," he hastened to say.

"No need for that. You young folks stay right here and

enjoy yourselves. Get acquainted," Pattison said in his large, hearty voice.

Seeing Tinette blush and lower her eyes in embarrassment, Earl moved quickly to put on his wraps. "No, I'll save that pleasure for tomorrow night, if I may. I believe you will be here, Miss O'Neil. Miss Mary has invited me to supper."

"Well," Tinette hesitated. "Some of the high school group planned to go bobsledding tomorrow evening —"

"Good!" interrupted Dr. Pattison. "After supper you can both hike out on your sleds. Be good for you." He was halfway up the stairs as he spoke. Then stopping to turn around he boomed, "Come early now so we can have supper over in time."

Earl Norris bade Miss O'Neil a formal good night and opened the door to leave. Then, as if on sudden impulse, he turned and stepped so close to Tinette that he almost whispered, "I'll hold you tight on the sled."

The next evening the four of them sat down to supper in a small alcove back of the sitting room. Young Dr. Norris seated the ladies. He took special care to press the back of his hand into Tinette's soft upper arm as he pushed her chair into place. Tinette's response was merely a stiffening of her back, though it would have been impossible not to realize that his action was intentional. She did not betray her inner confusion at such uncalled for intimacy, but she barely spoke throughout the meal. The two men carried on a lively conversation about the advisability of combining their offices into one. Mary Pattison, who at times received her father's patients, was delighted but Tinette had no interest in the plan.

At the close of the meal, Dr. Pattison brought up the subject of the evening's entertainment. "Go get your duds, Tinette, and I'll bring my old sled around for Earl. It's old, but it will seat you both."

"Well, Doc," Tinette said, "Bertha and Sylvia said they would walk by and I could go over with them."

"I'll tell them you've gone. Now scoot," Doc ended.

Earl was about to suggest waiting for Tinette's friends but the old gentleman did not give him a chance to speak. By the time Doc had showed Earl the sled on the back porch, Tinette had crossed the hall to the bedroom at the front of the house. She closed the door behind her and stood stiff in anger for a moment. Just as she was trying to decide on a future course of action she heard Bertha's and Sylvia's voices on the walk outside. Within a few minutes she joined them. The three girls and Dr. Norris trudged single file along the snowy path towards the high school hill where they would be sledding.

The evening's sledding proved to be a big success. Gaiety prevailed and Earl divided his attention among most of the girls there. He guided the old sled down the hill time and time again, with a different girl almost every time. His promise to Tinette did not come true as she sat behind him.

Around nine o'clock some of the parents carried hot cocoa and sandwiches to the shivering crowd, and soon after the refreshments were gone the crowd disbanded. Earl led the way home with Tinette, Bertha, and Sylvia following closely behind. At the Pattison home he bade Tinette another formal good night, left the sled on the porch, and walked the other girls to their home.

The next day and night passed quickly for Tinette as she helped Mary with the shopping and housework. After a light supper, the two of them frosted a cake and put things in readiness for Sunday. The coal oil lamps were extinguished early, and Tinette fell asleep easily.

The Pattison family had been among the earliest to settle in Marshall County. The original Pattison and his brother, uncles to the present Dr. Pattison, had preceded the O'Neils into the area by some seven or eight years. Neither brother lived very long but they had been vitally instrumental in the building of the small Episcopal chapel at the north end of Waterville. It was erected close to the cemetery where they now lay. The doctor, Hugo Pattison, obtained the uncles' property. He kept a live interest in the little chapel and seldom missed a Sunday service.

On this particular Sunday morning Doc led the way into the family pew, followed by his daughter and Miss O'Neil. Tinette was just leaving the seat to drop to the kneeler for prayer when Earl Norris brusquely took hold of her elbow and scooted her over enough to make room for himself. Mary gasped but moved close to her father. Tinette was surprised to see him and annoyed at his proximity. Throughout the service she had to keep moving her left arm to keep it away from the pressure of his arm. Later he walked beside her on the way to the Pattison home, assisting her at every street corner. It seemed to Tinette that he never missed an opportunity to touch her shoulder or arm or hand, and she hoped Miss Mary was not aware of it as she was acting stiffly as it was.

However, the dinner hour passed very pleasantly with a discussion of Lent and all it could mean to the soul.

Mary leaned toward Earl with an unaccustomed and slightly silly grin, "I think the idea for Wednesday night Litany services is a good one."

"Miss O'Neil." Earl spoke as soon as the subject was mentioned. "I shall see you to each Wednesday night service and home again. Gladly!"

"She lives in the country," Miss Pattison pointed out.

"Yes — " Tinette spoke in the same hesitating manner which had come upon her the first time this young man had spoken to her so boldly. "I do live quite a ways — "

"Oh, I know where you live. It is not far. Doc here pointed it out to me when I went with him on his rounds yesterday morning. It isn't even a mile!"

Pattison laughed and said, "You can get in practice by taking her home right now. Time you get the cutter she'll be ready to go."

Before Tinette could speak, or Mary protest, all of it was arranged. Goodbyes were said rather hastily and Earl put Tinette and her belongings in the cutter. The horse had hardly turned southward towards the O'Neil farm when Earl took Tinette's mittened hand in his own. He did not seem to notice her struggle to pull away, and soon the slender hand lay quietly within his sound grasp.

When this matter was settled he turned his attention to conversation.

"I liked sitting with you in church."

No answer.

"I could hear you recite but I couldn't hear you sing."

No answer.

"Don't you like to sing?"

Tinette shook her head.

"Don't like to talk either, do you?"

This brought a laugh to Tinette's lips because talking was her favorite pastime. Everyone was always charmed with Tinette's quick, cheery chatter and her choice of words. Earl laughed, too, and felt encouraged to go on.

"I love to hear you talk."

No answer.

"You know, this has been a very happy week for me."

"This is our place here," Tinette answered.

Earl went right on. "Have you divined why? Because I have decided that you are the girl I am going to marry. When you have made up your mind that I am right, you must tell me."

After he had lifted the startled girl and her belongings down from the cutter, he made one more statement before taking his leave, "Any time will be all right with me."

# CHAPTER EIGHT

## Suitors

Throughout the spring and the first few weeks of summer, Earl was Tinette's most constant escort. In some instances he was elegantly formal in his demeanor toward her. At other times he was boldly familiar in his attitude, and especially in his speech. But no matter how often he found occasion to press Tinette's hand or arm with his own, or even to brush past her pressing his chest against her shoulder, never once did he attempt a kiss or an embrace.

As for Tinette, she found her tongue and returned to her own normality of fast and cheery speech. However, speech did fail her whenever Earl spoke to her in what she considered bold familiarity, and she could not keep her body from straightening into rigidity at his touch. Sometimes she felt as if she would like to find herself in his arms being kissed by his handsome mouth. But such ideas came to her, as a rule, when she saw other girls making up to him. Mostly she liked for him to keep his distance.

As June was drawing to a close, Earl did a strange and surprising thing. As suddenly and as strongly and as constantly as he pressed his attention upon Tinette, he now withdrew it. At church he still sat in the Pattison pew, but beside Pattison himself. This seemed to gratify Miss Mary, who never missed church. However, she just as consistently never attended any social affairs, so she didn't know that now when Earl attended a social, he only spoke briefly to Tinette and allowed others to escort her home.

Tinette naturally felt embarrassment over this development and as a consequence stayed more often at home. She began to spend hours in the field with her father, helping with the shocking of wheat, or driving the hay rack. At other times she would find some shade near where the men were working, and she would sit and read.

One day as she sat leaning against a tree in the field, her father sat down to rest beside her. Taking off his straw hat he ran a hand through his thick black curls. "Hot enough to be summertime, ain't it, Honey?"

"Maybe that's why they call it July, Papa."

"Smart alecs, be they?" commented the fifty-nine-year-old farmer with his lips forming a downward grin.

"Here. Have a cool drink. Lemonade," Tinette offered.

"Don't care if I do. Some o' Mom's good cold lemonade."

"Mother is fixing some of her good hot fried chicken for dinner. That should make you do a good afternoon's work."

"I didn't figger on comin' back to the field this afternoon."

"Why not?" Tinette asked.

"I'm goin' callin'. That's why." When Tinnette made no reply he went on. "You know I told you someone had bought that land." He stopped to drink more lemonade and Tinette fell to thinking about "that land." Her father had referred to the piece of ground between their land and the main road. It was the plot he had wanted to own for over a quarter of a century.

"Has he moved in, Papa?"

"I hear he has. That's what I'm aimin' to see."

Later that same day Tinette was at the well drawing water when her father rode in on Pony. Calvin always kept a small, shapely horse for riding and he always named it Pony.

"Any luck, Papa?"

"No sight of him. George Root was ridin' out, said he'd seen him in the saloon. Old bat," he answered disgustedly.

"What is the owner's name?" Tinette asked.

"Bill Walton."

Late in July Tinette had spent Sunday afternoon in town with Bertha and Sylvia Bond. Having ridden over with a neighbor for the morning service, she had gone home with her girlfriends for dinner. About four o'clock she decided to walk home.

At the edge of town sat the house that had been involved in the recent trade, and it was part of this land that Calvin had wanted to purchase. As she got close to the property, Tinette noticed a buggy and a team in the yard. She was a person who had a natural curiosity about people, friend or stranger. She shifted her parasol to her left shoulder so that she could have full view of the buggy. In it sat a man of thirty-five or so years, perfectly motionless. If he had moved one way or the other, Miss O'Neil doubtless would have gone on about her own business. But when she saw that he remained immobile, she wondered where in the world his thoughts were carrying him. From the side he had a fairly nice appearance and Tinette could not help wondering what a full face view would show. She cautiously walked into the area in front of the house (it was too scraggly to be called a yard) and approached the buggy. Before he became aware of her she had time to notice the roughness of his face. It had more than a hint of the cynical.

"Jove," was his first word. "You don't by any chance go with this shamble, do you?" He added a jerk of his head in the direction of the shack of a house he had recently acquired.

"Why — uh — I've always thought that cabin — er — a — well, at least it has possibilities," Tinette stammered, ashamed of having gone so far into the privacy of the stranger's property and his thoughts.

Softening, his voice became almost merry as he said, "I'd be obliged to you if you would point them out to me, Miss — Miss — "

"O'Neil," Tinette replied, becoming more embarrassed by the moment. She began to wonder how she could get away. Should she apologize or merely say goodbye? She had a notion simply to turn without a word and leave him to his old house. It was a mess, but she knew through gossip that he had seen it before affecting the trade that made it his.

Tinette made a pretty picture that day with her blue umbrella framing her deep blue eyes. However, Mr. Walton scarcely had a chance to see them, for she kept them looking either at the house or down at the buggy wheel she was standing by.

"If you live close by, maybe you can be right useful in helping me," he said with a hope that she would look up at him.

Oh! Tinette thought. He is awful! Shaking her head ever so slightly, she turned and began to walk hurriedly out of the yard.

Grinning at her sudden departure, Bill Walton called out, "You going, Ma'am?"

Tinette pretended not to hear and she hastened up the steep hill which later became known as "Walton Hill." Above the top of the hill was a level space just long enough to allow one to catch his breath after the climb up the long hill. Then there rose a small incline which led on to a good piece of level ground. This piece Calvin had named the "Long Stretch." At the south end of Long Stretch was a draw, and under the culvert there passed a stream on its way from the Stanton farm to the O'Neils' pasture. Tinette had walked so fast that she hardly had time to think before she had reached the end of Long Stretch. Then she said aloud, "He really is awful! He didn't bother to get down from his rig or even tip his hat. Let him develop his own possibilities."

Before the end of July Tinette attended two parties in the Cottage Hill district south of their farm. She was quite surprised to see Bill Walton at both of them. He became properly introduced to her, and both times he managed to be coupled with her for refreshments. His manner was abrupt, and although he never used bold speech or intimate touch, such as Earl Norris had employed, at the same time there was a rude familiarity about him that repelled Tinette. After both parties he tried to insist that she ride home with him as her escort. Both times she stubbornly and successfully refused.

Bill had been in the community a few weeks, entering well into the spirit of things before Calvin O'Neil tried to approach him about the property. Walton was seated on his rickety old porch one day, whittling a piece of wood, when his neighbor rode in.

"I'm Calvin O'Neil, neighbor to the south," said Calvin, pointing in that direction.

Walton made no attempt to welcome his caller. Continuing with his whittling, he muttered, "I know it."

Sensing no friendliness, Calvin spoke immediately of his hope for a transaction. The proposition of buying the piece of land he so badly wanted carried such a splendid price that it gave Bill Walton an idea. He pushed his hat to the back of his head and looked with narrowed eyes into the distance. To himself he said, "That girl's plumb daffy about her old man, and the old codger seems to have his heart set on buying that property."

Then looking straight into the older man's bright blue eyes, he said, "Well, I ain't selling."

"You ain't sellin'?" Calvin repeated in disbelief.

"Might be you could get ahold of it though," the younger man offered with what resembled a leer.

"How?" Calvin asked.

"I made friends with your daughter," Walton said flatly.

"Well?" Calvin wished Bill Walton would get on with what he was trying to say.

"I'm not fixing to let it stop at friendship, old codger. Someday these farms will be under one name." Walton gave Calvin a knowing look.

Calvin stared at Walton in total disbelief. He knew that Tinette would never give herself to such a crude scoundrel, and he began to put his thoughts into words: "You won't — you can't — Tinette wouldn't — "

Bill Walton misconstrued Calvin's words to apply to an immediate transaction and he answered his prospective father-in-law with an insulting spit towards the ground at his feet. "Nope. We'll wait awhile."

Before Calvin O'Neil had left his neighbor's place, he made up his mind to speak to his old friend, Dr. Pattison. As he turned Pony toward town he had a mixture of feelings. He was bitterly disappointed at once more confronting an adamant owner of the property he needed in order to complete his farm and have an outlet to the main road. It seemed like the last chance. Along with his feelings of disappointment was his intense dislike of Bill Walton. The man gave Calvin a sense of loathing disgust and uncleanness.

Calvin reached the town office just as Dr. Pattison was setting out on a call.

"Come to see me, Cal?" the physician called out in a friendly voice as he stepped into his buggy.

"In a hurry, Doc?" Calvin asked, bracing his arm against the rig.

"What's ailing you, Cal? Something wrong?" Doc

asked, his eyes looking concerned for his good friend.

Calvin O'Neil nodded and said, "You'd best be movin' on, Doc. I'll untie your assistant." It was Calvin's usual jest but the twinkle was missing from his eyes. As the doctor picked up the reins he leaned towards the worried man and asked again, "What's ailing you? Is the Walton fellow bothering you or Tinette?"

Calvin nodded, "What's the matter with that lazy dentist? Looks like he could cut out that Bill Walton. Don't he want to?"

"Now don't you worry, Cal," his friend spoke with confidence. "Earl will cut out Walton before the week is done."

For the remainder of the summer Tinette had the attention of both young men. Earl Norris courted her with all the grace and poise he possessed, and most of the time she was glad to accept his offers as escort. But during a party Bill always contrived to secure her for a partner enough to make everyone think that he was very much in the running.

Tinette began to feel as buffeted about as a ball of tumbleweed in a prairie windstorm. At night she tossed in her bed as she pictured herself a sheep running from one side of the corral to the other, trying to escape capture.

One night the Cottage Hill group invited some of the Waterville young people out to a watermelon party. Tinette rode out early with Bertha and Sylvia Bond in order to help with the plans for the evening's entertainment.

The party was in full swing before either Earl or Bill made an appearance. It looked as if one or the other had been waiting outside to see if the other one came. As soon as the men entered the hall, Bill immediately proceeded to leer at her wherever she went. Earl noticed at first glance how well her new dress showed off her full figure. The dress was pale green muslin and was cut so that every curve was in evidence. As she leaned over to pour some lemonade, the soft material let her ample bosom gently sway forward. Earl walked over to her and ran his hand up the sleeve which fell loose from her elbow.

"Earl," she said with a quick step away from him, "you must not do that."

"It's as close as I can come to what I really want to touch," he whispered as he fastened his gaze on her bust.

She felt his gaze and realized his meaning with a mingled sensation of half-fright and outright disgust. Such thoughts were so foreign to her own that in her innocence they seemed obscene. Turning away as quickly as she could regain her self control, Tinette hastily stepped outside where most of the people were enjoying watermelon.

Not wanting to have her state of confusion discovered, she stepped back into the shadow of the house until she could regain her composure. She instantly burst into tears as the hoarse voice of Bill Walton whispered in her ear, "Let's see if you feel as good as you look."

Later, after an abrupt departure from the party, Sylvia and Bertha deposited Tinette at the front door of the O'Neil farmhouse. The girl ran into the house without a backward glance or a final word to her friends. Unable to hold back the tears, she ran through the hall to the sitting room and flung herself upon her small bed. Within a few minutes her father had pulled on some pants and appeared at her side.

"Tinette, Honey," he whispered as he stepped to her bedside. "What's the trouble?"

"Oh, Papa," she moaned.

"Bill been after you, Tinette?"

"Papa," she sat up, trembling beside him. "Do you still have that address of the man in Kansas City that's your kin? Do you?" Her anxiety was evident in her voice.

"Reckon so. Why?"

"You said maybe sometime you would travel over and see him, didn't you?" Tinette continued.

"Figgered I might, sometime when I had to go there with grain or for some matter. Been a long time since I seen any of my kin, even if this one is only a distant relative that I've never seen or heard of."

"Papa, please go, and take me. Please take me," she begged as her tears increased.

Calvin ran one hand over his short black curls and spoke softly, almost to himself. "Well, the summer work's by no means over, but mebbe Sam could handle it fer me. When do you aim us to go?"

Her father had spoken in a low voice but Tinette heard him and she looked up with hope.

"Right away. Tomorrow. Tonight." Her voice rose in pitch with each word and then she began crying in relief.

Nancy O'Neil had been standing at the door quietly watching and listening. She felt sympathy for her daughter, whatever the cause of her tears. She went back to her bed noiselessly, shedding a few tears herself, but happy that Calvin had agreed to Tinette's pitiful plea.

The letter to which Tinette had referred was from a distant relative who wrote that he had recently left Pittsburgh to make his home in Kansas City. He was employed there by a broker's firm. He had obtained Calvin's address from his great aunt Anna to whom Calvin had written after the death of his father. In closing he had urged his kinsman to pay him a visit. He had signed the letter "Weston O'Neil."

After Tinette's "siege of nerves" as her mother called her frenzied behavior, the O'Neils made rapid plans for departure.

When they arrived in Kansas City, Calvin took rooms

for the three of them at the Coates House. He got in touch with the cousin through the firm of Dermott and Jones. Arrangements were made for him to join them for supper in a private dining room at the hotel.

The twenty-six-year-old man arrived punctually. The O'Neils were waiting for him in the big lobby. As Weston O'Neil entered the building he removed his hat exposing a head of short black curls which easily identified him as the relative they were expecting. Except for less distinguished eyes and a smaller mustache, Weston was a counterpart to the O'Neil who rose to greet him; a less handsome but more striking version of the older man.

"Calvin O'Neil," said the farmer.

"Weston," replied the other. "And I'm proud to know you, Sir."

The two men shook hands in a warmly sincere way that seemed to indicate how bereft each had been in his complete isolation from any kin.

Calvin spoke again. "Here, Wes, I want you to meet my wife and my daughter. This here's Tinette and her mom."

"Mrs. O'Neil," Weston responded as he took her tiny, workworn hand in his. "Let me thank you for bringing this man along to see me. The O'Neils have been separated long enough."

Nancy O'Neil smiled up into the friendly blue eyes, and what she was thinking would have startled Calvin as such a thought had never been expressed: "This is the son I have dreamed about. He would have looked like this boy. He would have been like this kind, sweet boy."

Something in Nancy's mothering look went to Weston's heart, and the two stood an extra moment searching one another's face. Calvin interrupted the silence by touching the young man's arm.

"Meet our girl, Tinette," Calvin said.

Weston bowed formally and took Tinette's hand.

"You look like my father," she said in delight. "You are the only person I have ever seen who even slightly resembles him."

"He's the first kin you've ever seen, too, Honey," her father reminded her.

"But people have a habit of looking alike in one way or another," Tinette answered. "But no one ever dared to have Papa's eyes or hair or nose."

"I thank you, Ma'am," Weston smiled, "because I think he would be a fine man to take after in any respect. But let me tell you that it is plain that you are his daughter. So someone else resembles him, too." Weston beamed, then turned to Nancy, "Don't you think so, Mrs. O'Neil?"

Nancy O'Neil was absolutely enthralled, and absentmindedly nodded her head.

"Well, let's go on in to feed," Calvin suggested, turning to lead the way to the dining room. At the table Weston carefully seated Nancy and Tinette, paying special attention to the older woman.

"Are you comfortable there by the window, Mrs. O'Neil?" Then leaning over to put his hand on hers, he added, "You make a pretty picture in front of that dark drape with your white hair and pink dress."

Nancy smiled indulgently at the young man. With the birth of Tinette, Calvin had forgotten about his former yearnings for a big family. His understanding that there could be no more children and the joy and admiration he felt for his daughter ended his previous longings. Calvin certainly never knew that his wife entertained any thought at all about another child. But now and then during Tinette's twenty years, Nancy had indeed envisioned what a son might have been. Now she felt that Weston was the embodiment of her secret creation.

When the large array of food was placed before them,

Weston commented, "I suppose you are the one responsible for this fine selection, Miss O'Neil?"

"Oh, no," Tinette replied in her usual frank manner. "I suppose that Papa decided on the menu."

"Nope. Jest told 'em to give us good fare," Calvin said.

Having complimented the ladies and making them feel comfortable, Weston turned his attention to the man of the family. The ladies studied his grace and poise. They were absolutely oblivious to any faults he might possess.

"Now, Calvin," he began. "If I understood Aunt Anna right, she was your aunt by marriage."

"My father's aunt by marriage," Calvin corrected, "but what do you mean, she 'was'? Did she die?"

"I thought that you knew. It has only been a short time since I heard, so perhaps by the time you get back home, you will have received the word."

"No, no. There hasn't been any communication between us," Calvin said. "No love lost either. But I shouldn't say that to you."

"My O'Neil grandfather was a cousin to her, a double cousin, I was told. The relationship was a long way back but Aunt Anna was a peach to me while she lived."

"How did you happen to come to Kansas City? Where did you hear of the firm you're with, uh, Dermott and Jones?"

"I just came here by chance. I had heard of Mr. Dermott in Pittsburgh. He did business there and was respected. I feel lucky that he actually gave me work."

"What work do you do for him?" Calvin asked.

"It's called broker work, Calvin," Weston explained.

"Like it?" Calvin asked skeptically.

The young man nodded. "It is interesting and very fine people to deal with." Turning to Nancy he winked and smiled, "Oh, there may be a thing or two under cover, so to speak, now and then." Looking again at Calvin he became very serious and said, "Very fine people. They've done a lot for me."

Weston's manner appealed to both women. Both thought that he was fascinating and good-looking. Even Calvin felt that Weston inspired a certain trust, or perhaps goodwill. Later Calvin remarked to Nancy that the young man did not seem to be a braggart, although he was apparently doing quite well for himself.

Before Weston O'Neil left the hotel that night, he had been granted permission to call for his relatives on the following night, insisting that it was his turn to be the host.

The next evening he met his relatives at the Coates Hotel. Tinette had dressed with a degree of hesitation. She wanted to and yet felt reluctant to wear her best and newest summer dress, the pale green muslin which seemed to have been the cause for the ungentlemanly remarks she had received the first and last time she had worn it. Her mother was completely ready, wearing a mauve dress of rather sheer material, one that Tinette had made for her in midsummer. Nancy urged her daughter to put on the green dress and admonished her not to keep the young man waiting. Upon Weston's arrival he found both ladies prettily and tastefully attired.

"Good evening," he spoke in his pleasant voice as he stepped into the lobby and removed his hat.

"Howdy," was Calvin's hearty reply. "Ladies here all spruced up fer you."

"They both are as pretty as any ladies in this old city tonight," Weston beamed. Addressing the women, he added with a slight bow, "It's proud you'll make me to come along now."

"Come on, Nan. Let's not tarry here a gabbin'," Calvin said as he preceded her to the sidewalk. Weston offered Mrs. O'Neil his arm, and Tinette ran past them to walk with her father.

"Would you feel like walking a bit, Mrs. O'Neil?" Weston asked. At her acquiescent nod he raised his voice to speak to the others. "Let us go right down on this walkway. On north a ways is a fine place to eat."

After a pleasant supper Weston hired a hack to take them to the Missouri River where a showboat was moored. The O'Neils were entranced with the boat itself, and later with the show. Weston sat behind the two women, and although he was courteously attentive to Nancy, he found himself wanting to look more and more at Tinette. He was delighted with her naivete, her ready laugh and her sincere appreciation of anything done for her pleasure. He thought she made a lovely picture with her slightly auburn hair and her clear skin above the pale green of the dress. In fact, Weston decided that with her coloring, her lovely figure, and her expressive eyes, Tinette was a real beauty.

Even Tinette's mannerisms charmed Weston, especially her frequent habit of gently passing her hands over the hair above each ear. As she did this, her full sleeve fell back to the elbow revealing a soft patch of white skin in the inner bend of her arm.

Tinette was pleased with Weston for his tender behavior toward her mother. And she was gratified to see that he and her father had a mutual admiration. They were perfectly natural and at ease in each other's presence. She began also to be aware of his interest in her. She instinctively liked him and enjoyed entering the scope of his attention.

The young man ordered a hack for the ride back to the hotel. Weston and Mrs. O'Neil sat across from Calvin and Tinette. Weston observed that Tinette became quite animated as she recalled bits of the show she had enjoyed. Suddenly she said, "I would like to see that again."

"It will have to be tomorrow night," Weston answered

immediately, "as the boat leaves on the next day."

"All right. Tomorrow night it is," Tinette said emphatically.

"Not fer me," Calvin said. "How about you, Mom?"

"Not for me," Nancy echoed. "I intend to spend some time in those sidewalk rockers. Otherwise I'll never be rested enough to make the trip home."

The next few days Weston and Tinette saw a great deal of one another.

At first he only called in the evening, but after a couple of days he hurried over to the hotel as soon as he could leave his work.

Calvin had noticed that Tinette had begun to chatter like a monkey and had taken on a gay frivolity that was pleasing to Weston. He seemed charmed by her in every respect.

Then Calvin noticed a change, the living-in-heaven period had ended partially. Only Tinette appeared to still be up in the clouds. Weston seemed to be puzzling over something, as if he had serious thinking to do but was trying to avoid it.

One morning the young man showed up at the Coates House early enough to find Calvin alone in the lobby before the women had left their rooms.

"Calvin," he said anxiously, "where can we sit and talk alone?"

"Why, right over there will be the best, I reckon, Wes." After they were seated in one corner of the room Calvin asked, "What in the world is the matter, Boy?"

"Calvin, there is a girl in Pittsburgh. She is the reason that I had to get out. She is the reason that I came West."

"Well," Calvin waited.

"I detest her," Weston spoke with a trace of agony in his voice.

"Well?" Calvin asked once again. "Who is she?"

"She is Mary O'Neil. She is my wife."

Calvin was stunned. Already he had wanted this boy for his own to take home. The love between him and Tinette had been obvious. Even Nancy had surprised him once and referred to him as "our boy." He sat forward, letting one arm rest on his knees while the other hand roamed restlessly over his tight black curls.

Weston leaned close to explain, "Mother insisted on the marriage. Mary was the only girl she would ever think of allowing me to see. On Mother's deathbed she made me promise. I had to fulfill my vow and marry her."

Calvin glanced up at Weston and caught a look of bitterness on Weston's usually kind face. He said, "Talk on, Boy."

"I stood it for over a year," Weston continued sorrowfully. "She whined. She ailed constantly. Never could I count on a cheerful or helpful word from her. She is a selfish, whining fiend. I left her."

"She know where you went?" Calvin asked.

Weston shook his head. "No, but she will get along without me. I guess the reason that Mother was so insistent was because of her money."

Both men sat in silence, feeling clearly the misery of the other. Finally Calvin touched Weston's knee and quietly said, "One fact is certain. She is your wife."

Weston looked up sharply and said, "One other fact is certain, too. Tinette and I love each other."

Calvin O'Neil groaned, then sat silently staring at the floor a moment before slowly rising to his feet. "Shall I send Tinette down?" he asked, feeling much older than his fifty-nine years.

Weston nodded, then finding himself alone he went in search of a secluded alcove where he could talk to Tinette. She came down immediately. From her father's expression she knew that something was wrong, and as she ran towards Weston she asked the question, "What has happened?"

Weston took her in his arms, kissing her hair. "Tinette," he said, "you have become the dearest thing in my life and I love you. But I am not a worthy man. I – I have to give you up."

Tinette loosened his grasp and looked up into the face she had come to love, "Weston, what do you mean? Why do you have to give me up? No! I will not let you give me up. I love you."

She strained on tiptoe toward him and he allowed himself one long and fervent kiss before he said it: "Tinette, I am married."

The girl in his arms uttered a cry, then sagged against him. Weston held her close and stroked her head again and again as painfully he told her the story her father had heard earlier. Minutes slowly ticked by as Tinette poured out her grief in tears. Again and again they kissed and through tears avowed their love.

Suddenly Tinette pulled out of his arms. Looking into his eyes she said, "No matter what has happened or will happen, I love you. Never shall I love another." She placed a light kiss on his trembling lips and ran quickly from the room.

Calvin had stayed in the hotel room in order to break the shattering news to Nancy. She barely had time to compose herself before the door opened and Tinette entered, weeping. She wept for the plight of her loved one as much as for her own. She wept disconsolately.

Finally, the tears stopped and Tinette grew pensive. That day their vacation ended. Before they reached the farm, Tinette O'Neil had planned her future, putting away that which was past.

When the O'Neils returned from their trip to Kansas City, the stock and food show was already in session at the

## SUITORS

neighboring town of Blue Rapids. They urged Tinette to attend at least one day since that had been her habit for several years. She agreed with a shrug of the shoulder, and on the closing day of the fair she and her father rode over with the Stanton boys. The men were busy with the stock so Tinette rather aimlessly explored the food area. Finally, she went to sit in the wagon to wait for her father. All at once she was surprised to see Bill Walton climb up into the wagon and sit down beside her.

"Bill!" she gasped. "You must not sit here with me alone."

But Bill had been watching for the moment when she would leave the grounds to look for her father, and perhaps wait in the deserted area of the carriage. Hastening to take opportunity of the situation, he slipped one arm tightly around Tinette's shoulder and the other about her waist. While struggling to free herself, she heard his proposal of marriage. It was really more of an edict than a proposal. Bill vowed that she was his, and that by winter they would be living in a new house on the coveted eighty acres. Then he dared to kiss her. With that kiss all the fury which his crude nature had been engendering in Tinette's heart broke loose. She jerked away and jumped to the ground before he knew what she was doing.

She took time to give him a loathsome look before running back to the fairgrounds. Her last words to him were, "How I hate you, you will never know."

It took a while for Tinette to lose her nervousness and to regain her former composure. It had been a trying summer for her. She had fallen deeply in love with one young man, rejected another, and decided to marry still another.

## CHAPTER NINE

### A Rocky Marriage

**B**efore Tinette O'Neil married Earl Norris, she had indeed planned her future. Having decided to accept on some future day the young dentist's proposal of marriage, or edict as he had issued it, she began to train her mind to view the decision logically.

First in order came a disparagement of the character of Weston O'Neil. She must not look at him as desirably handsome; she must view him impersonally as a man she had merely met. A man who had come into view and then faded out with features becoming so indistinct that they were unrecallable.

(Oh Weston! I see you as clearly today as if you were standing beside me. Your blue eyes, that earnest look. A closed-mouth smile, an open-mouth grin. I close my own eyes and feel my fingers touching your face. Caressing your face as your arm draws me to you. My love!)

She must not think of him as a man so stalwart as to inspire respect; she must see him as a self-promoter, a man of devious actions. Dishonest, a pretender, a cheat. Had he not come a-wooing, knowing himself to be a married man? Did he not present himself as a man alone in the world, free to squire an innocent girl — herself — here and there around the city?

(Oh Weston! You had been forced into a hateful marriage, and you had left that misery and gone away to forget it. And did I not see you draw yourself away from me in an effort to be loyal to that spiteful oath which had been forced upon you?)

As the days went by, Tinette stayed at home, yet to herself as much as possible. She would go to the orchard, now bare of all fruit, and virtually hide among the branches to weep and sob and talk to herself.

(Had he not gone to her father as soon as he realized the damaging impasse into which their love was leading them? Yes. And why had he not come first to her instead of seeking the strength of her father's character? That was it — he was weak. Weston was weak. He could not face up to his ordeal alone. By his own admittance, he had been his mother's pawn. He had allowed her to rule him and it was weak submission which had led him into that wretched marriage. Yes, he was weak. Unmanly.)

She would take the well-worn path which led almost imperceptibly downward towards the creek which ran through the sheep pasture. Earlier in the summer, and in happier summers, she had rested here among the violets. Mama, she thought, always loves to see me coming back to the house with a cluster of violets. Nosegays, she calls them. Yes, and did not Mama's eyes light up in the same way to see Weston coming back to the hotel for one more and still one more visit and get-together?

(How dared he, Tinette whispered to the memory. How dared he extend such rare happiness to Mama when he had no right to do so! Mama loved you, Weston. She would have brought you home and nourished you with her love. She expected to. And you knew from the start that there was a barbed fence around you that was going to cut anyone who came close. How dared you do such a thing to my defenseless mother? You saw that she loved you. Yet you allowed and encouraged her feelings to grow. Weston O'Neil, how dared you cut Mama — and me — with that barbed fence of which only you were aware? You are an evil man. Your heart is as black as your hair. You hurt my little mother. I detest you!)

Next came her wish to live as she had before her twentieth year. No suitors in her life. No Earl Norris, no Bill Walton, and certainly no Weston O'Neil. She would take a huge imaginary bar of mother's castile soap, and with it she would scrub away the blight of suitors.

There came to her mind the old schoolhouse chalkboards in which would be imprinted a beautiful design covered over with colored chalk. With eyes closed she could see herself washing and scrubbing away at the blackboard picture which had changed from a scene of nature to a likeness of Weston O'Neil. The board became messy as the picture blurred. The colors ran together, distorting the face. Determined to rub it out, she squeezed her eyes tightly together as she fancied herself applying "elbow grease" to the task. For some time the scene kept returning to the imaginary board, but gradually it began to fade and finally was obliterated.

Tinette felt drawn to her mother more than ever before in her life. Where once she had spent most of her waking hours outside, reading under a tree or following her father from task to chore or riding her pony around the farm, she now preferred to stay inside with Nancy. She helped with the indoor chores in a more thorough way; where previously she had given her tasks a "lick and a promise" in order to escape outdoors. It was at this time that she learned from Nancy the secrets of being a good cook.

Several years earlier Nancy had hung a fancy quilt on the north wall of the sitting room and placed beside it a narrow bed for Tinette. This had been done at the time Calvin had built a hall leading out of the sitting room. The front door was removed thereby from the sitting room and replaced at the north end of the wide, new hall. To the east of the hall Calvin built a large square room to which Nancy moved their bedroom furniture. In time the old bedroom had become a spacious dining room used for company dinners at noon every Sunday. To the west of the hall he built a large square parlor, the only door of which opened onto the hall. The main extravagance of the two new rooms had been the six fine windows claimed by each. Neither had a fireplace.

There had been several times when Tinette had chafed at having her bed in the sitting room. Once it had been ideal. She had thought herself a big girl when she was given permission to have her cot-size bed moved from her parents' bedroom into the nearly empty sitting room. And even after the sitting room had been furnished with chairs in the bay window area and chairs near the library table which stood in the center of the room and with three rockers close to the fireplace, even then Tinette had not felt an incongruity in having one wall devoted to bedroom appurtenances.

Later, however, when she was in her later teen years, the wonderment of the situation came to her. Why such an arrangement had ever been devised was her question. Why had not the parlor, scarcely furnished for awhile, been her bedroom instead? Upon puzzling over it one night in her sitting room bed, there came to her sort of a memory of her father's voice, suggesting to Nancy that Tinette should surely occupy one of the new rooms. No memory of her mother's reply came to her. After the situation became nettlesome to her, she asked her father why she didn't have a room of her own. He only said, "Your Ma wanted a parlor."

After the shock of losing what she looked upon as the love of her life, Tinette's feelings about the bed changed. She was glad that it was close to her parents' bed. Just a wall with a beautifully designed quilt separated her from her mother. Before falling to sleep at night, she would snuggle close to the wall, holding an edge of the hanging.

Last in the process of shaping her future came Tinette's determination to see Earl Norris as an acceptable and even desirable partner. Her parents never once mentioned Weston. On the other hand, neither did they speak of Earl Norris. All of the wedding preparations, all of the courting, all of the unfolding decision that Earl's character was sterling, all of this was only in Tinette's mind. Therefore, it came as a great surprise to the O'Neils when Tinette announced to them the date on which she intended to become Mrs. Earl Norris.

It had happened that one day Nancy had run short of nutmeg and Tinette walked into town to get some. Her fate had arranged for Earl to be on the street as she entered the store. A former mayor, Mr. Frahm, had been explaining an insurance deal of some kind to Earl who could never recall later just how it had ever interested him; but the young man's interest flagged noticeably and the men parted. When Tinette left the store and reappeared on the street, she found an escort at hand.

"Miss O'Neil, I do believe?"

Tinette smilingly offered her hand, and leaving it in his for a moment, she looked into his eyes. A questioning look, he thought. Or perhaps just surprise at seeing him. No, it was more like an interested look. Earl dropped her hand, tipped his hat, and took her arm as he accompanied her down the street.

"I've missed seeing you."

She nodded slightly and smiled.

"I can't forget the first time I saw you. It was winter."

She gave several fast nods of her head.

"It was winter and I walked you home."

Nod, nod, nod in rapid succession.

"I told you that day to get used to the idea of becoming my wife."

Tinette became tense. He could feel it under his hand

which he had placed lightly on her shoulder. Her smile faded and her head did not nod. The young man thought that perhaps he had gone too far, but decided to keep trying.

"I can't walk you home today as I have promised to pull Peterson's tooth. But my offer still holds. Let me know the date."

The determined young lady clicked her tongue once, then inclined her pink face so that her poke bonnet shaded her blue eyes, and said distinctly, "Christmas Eve will do."

When Calvin O'Neil lifted the roof from the front of his house in order to build two upstairs rooms, he had no idea that a large family of children would in time come to inhabit the farm. The bride and groom occupied the bedroom above the room shared by Calvin and Nancy. Within four years three children cried, yelled, and laughed in the room across the hall. Ted was the oldest, and it was he who taught Sara, born one year later, and Franklin, who kept up the one-year-apart tradition, how to slide down the bannister. It was not a spiral, but it did curve handsomely to give an exciting ride to those who were in a hurry to reach the downstairs hall in time to race back up for a repeat cascade of laughs and thrills.

Nancy Hawkins O'Neil had not prepared Tinette for marriage. This was in the true Hawkins tradition and seemed the correct omission to Nancy. Farm life had prepared the way crudely and inadequately for the nuptial arrangement, but concerning the hazards of connubial living, enlightenment had been left to old Mister Experience. No psychology classes, no warning from old hands, no lectures as to the nature of the relationship.

Tinette was bewildered. Couldn't Earl understand that she did not want their marriage to be built upon embraces? Didn't he realize that ardor cannot be deeply appreciated when it was so constant? The closing of their bedroom door meant only one thing to Earl, familiarity with his wife. He would kiss her, hold her in his arms, undo her braided hair, assist her as she took her basin bath, and once they had gone to bed he would not release his hold on her until morning.

When little Ted was born, his mother almost worshipped him because the fact of his being had somewhat released her from the pressures of too much ardency.

"I have to nurse the baby," she would say. "Little Ted needs me." "Hush! You'll awaken Ted." "I am too tired to hold the baby, so I'll lay him in bed with us."

Conjugal duty prevailed, but by the time Sara, then Franklin, had arrived, Earl had begun to yield to a less zealous love. Three cribs in the room plus three million demands had a cooling effect.

Gradually, however, Tinette began to enjoy her husband's company. They learned how to speak with one another.

"How did the children behave today?" "All three napped at one time and I said 'Glory!' " "What else did you say?" "Hallelujah!"

Calvin built a pen when spring greened and put it in the side yard where shade fell from the pine tree. There the three children could play safely. Ted could use an hour in rolling a large, red ball to Franklin. Sara, born bossy and picky, would show the baby how to roll the ball back with his hands, frowning each time he rose on all fours in order to propel the ball with his head. Tinette sat in the yard swing to read, and when Earl returned from town she would relate to him something she had read. Maybe from "Kenilworth" or from Godey.

They began to arrange the empty bedroom with a bed for the boys, and a couple of low chairs beside a table. The three cribs remained where they were but, only Sara still slept in one.

"Do you want to braid my hair tonight?" Tinette whispered to Earl. He jerked his head toward the crib as if to ask, "What about her?" Tinette merely put a hushing finger to the lips in invitation to him.

About two-and-a-half years of tranquil living passed by for the young people. When around the older couple they remained a bit detached from one another. But when alone, their exchange was very pleasant, even lively. The children were happy and affectionate with their grandparents, and somewhat wild in their play. About this time a certain realization began to enter Calvin's mind, a certainty that the house should have more rooms. Up came the roof from the back part of the house, and more rooms took shape. One over the sitting room with a narrow hall between it and the one that was built over the dining room. At the back was added a storeroom.

Once more the boys were moved, this time to the new room directly behind their parents' room. A door connected the two rooms, and if Tinette wanted to leave it open she could watch the little fellows racing around the room or leaping into their bed. Sara now moved into the front bedroom across from her parents, and the cribs were rolled to the storeroom.

A new decade had begun for Tinette. Dreams of Weston had plagued her for the first ten years of her marriage, but sometime after the births of the three children, the dreams had ceased. Her husband became a real person to her. She was interested in what he had to say. She was proud of his dashing appearance and took to going out with him regularly. Calvin and Nancy rarely went to church, but on occasion Tinette would accompany Earl to services or to social functions.

One thing did surprise Tinette, and that was Earl's

attendance on the ladies. His one close friend, Doc Pattison, had died, and Earl seemed not to have formed any other close friendship. Tinette did learn that Earl was accustomed to spending part of most days at the old doctor's house, apparently in the company of Miss Mary. Tinette would not have thought much of that had she not received a rebuff herself from her former teacher. One day when she drove her pony cart into the village, little Sara at her side, she decided to pay a much overdue visit at the Pattison house. As she tied the pony to the iron post, she clearly saw Miss Mary's form appear at a window. Yet no matter how strongly she and Sara rapped on the door, it was never opened to them. Not a sound was made behind the door, not a stir or a flutter.

As the pair gave up and drove away, Sara clicked her tongue once, much as her mother frequently did, and said, "Daddy at least should have been there at the door."

Tinette was too bewildered by Mary Pattison's strange behavior to realize the import of Sara's remark. A bit later when she tried to think exactly how Sara had worded that sentence, it would not come to her. The child probably had meant that her daddy should have been with them, Tinette reasoned. Nevertheless, she did not see him in town. Quickly, being hurt and flustered by her former teacher's actions, she picked up a few articles in the dry goods store and headed for home. The nauseating sickness of a current pregnancy had just come upon her and she neglected to mention anything about her visit when Earl returned that evening.

During the next few months, Tinette Norris was a very sick woman. For the first one of her pregnancies she suffered stomach sickness and kidney pain and a terrifying dizziness if she barely rose to her feet. Her weight gain was tremendous, although only burned toast, soda crackers, or crisply fried bacon could settle her stomach.

Earl was punctual in returning from town each day, although Calvin wondered what activity lured him to leave home each day because a new dentist had virtually all of the town's business. Only a few oldsters who were used to Earl still employed him.

Earl walked into town each morning, and back each evening. He barely took time to eat, spending his evening hours beside Tinette, placing cool cloths on her head and patting her wrists with alcohol, or changing the position of her legs by elevating her feet on a stool in front of her chair or on pillows if she were in bed. Calvin one evening complained to Nancy that he never had a chance to talk to Tinette. Having only the evenings free and indoors, he invariably encountered Earl at her side giving tremendous effort to divert her from feeling so bad.

"Guilt," responded Nancy.

Calvin and Nancy and Sam Stanton from across the road looked after the children. Sara and Franklin played cards, read books, and more often than not played school, pleasurably totting up figures and making account books. Nancy could rely upon both of them for errands and for inside work. They helped dust and sweep and were especially good about working in the kitchen for short periods of time. As they did dishes they would play word games, describing countries or foods for the other to guess.

Sam Stanton often had little Sara and Franklin with him. Sam was the hired man and best friend, and was loyally devoted to Nancy O'Neil. As he worked around the O'Neil yards (his work never carried him into Calvin's fields, only to the corral or maybe up to the pig sty), only the children would try to keep up. Tinette allowed bare feet only on the lawn east of the house, so in their red sandals the children would scamper after Sam, finding places to sit near him while he worked. On occasion when his self-alotted work for the O'Neils was finished, Sam would put the children in the wheelbarrow and push them over to his farm to play.

Ted, the oldest, and his grandad were together for most of the daylight hours. There was no dawn that Ted could sleep through. As soon as Calvin was dressed for the day's work, down the steps Ted would tiptoe to join him. He was barely twelve, but capable as any hand Calvin had ever hired. Calvin might curl his lip mentally at the thought of Earl, but never in a million years would his manner reveal anything but acceptance of his son-in-law. The bitter disappointment he felt in Earl was many times overbalanced by the loyal love reciprocated between his daughter and his grandson and himself.

At the end of her pregnancy, it took days for Tinette finally to release the child from her ill body. The torture incurred by the birth was as keen as if a huge many-faceted machine were cutting its way to air. Earl seldom left his wife's side and never was absent from the room. He rubbed her with alcohol, he poured whiskey down her throat, he bathed her with soft, moist cloths provided by a weeping Nancy. Finally, Calvin walked to town in a rapid frenzy, and looking up a doctor who had only recently moved to Waterville, demanded help. But Doctor Thatcher was a recent bridegroom and had already promised his bride to take her to Blue Rapids where some of her Topeka kinsmen were visiting. Calvin ran nearly all the way home to tell Nancy that the young doctor would be out inside of an hour. Maybe two hours.

Tinette screamed out and Nancy ran silently weeping up the stairs. Before long the baby forced its way out — a tiny, fragile girl baby, so weak that her cry could hardly be heard before she was lost in death. The situation seemed unreal to the helpless mother, indeed she seemed not to comprehend it at first. Later she realized that all during the birth process no one could have helped. Her mother was as helpless in the situation as she was herself. No doctor could have saved the child. Except for a spate of tears now and then, she accepted the death as inevitable.

Tinette looked adoringly at Earl. He had stayed at her side constantly. He had devotedly reassured her. He had tenderly caressed and held her. He had given quiet assistance to Nancy. Tinette felt that there had been two births, one for the baby, though she died, and one of a new love between the baby's parents. Lying in her bed for a thorough convalescence, she could hardly wait for Earl to return from his day in town. She lay with her hand in his, and during the night with one arm around him.

In the year that followed, Tinette rejoiced in her renewed strength and happiness. Although she was more plump than she had ever been, she was shapely and her skin was rosy and glowing again with health. She gloried in Earl's strengthened love.

All year Tinette stayed very close to home, enjoying some periodic excursions in the carriage with Calvin. Calvin would lift Sara into the carriage beside Tinette, and allow the boys to ride backwards on the carriage. Then he would drive them here and there around the farm. Up this road to get a melon; over yonder for the Christmas tree; down the road to a lane which led to the pumpkin patch. Glancing at Tinette's glowing face, Calvin would be reminded of his lovely young daughter as she was before life hit. How would she take this next hit, he wondered.

Sara, coming back from an errand to town one day, asked, "Granmama, why is Daddy so much at the Pattison place?" Tinette had heard the question but not the reply because it had consisted only of a look of warning from Nancy to Sara, and a shushing finger placed on the little girl's lips. Sara had dropped a flower in the front hall, and her mother sank down onto the carpet to get it.

Why indeed, Tinette wondered. In a moment she had

realized that it was true. But why? Miss Mary must be a full ten years older than Earl. Was he such a baby that he needed a mother to love him? Then it was true that when Earl had recently made an excuse to travel to Atchison, Mary Pattison had met him there. Tinette had heard Mr. Kieffer tell Calvin one day that he had driven Miss Pattison to Blue Rapids when she could have caught the train right at home.

"Train?" asked Calvin, who along with Tinette knew that Earl had taken that very train.

"Yup," said the blacksmith. "I seen her myself a-steppin' up onto that train. Same one goes through here."

Tinette's love which had filled her with happiness and had blinded her to obvious infidelity, snapped and departed. Love departed. She felt her heart harden. Once before she had gone through a love loss, but this was different. This had been love for a husband, a love she had a right to, a love that had enriched her family. Family! What must she do? She would protect them. They would not be allowed to witness her disillusionment. She knew that she could not again give any affection to Earl, but she determined to so control her manner as to create a sense of harmony for the children and her parents.

Slowly Tinette stood and walked to the front door. She pulled aside the heavily embroidered curtain from the door's window and gazed out into the scene beyond. Her glance skipped the immediate front yard, the driveway, and the pear orchard. She was looking farther on, to the creek with its bordering violets and smooth rocks, a view that could be glimpsed only in her mind's eye. Memory showed her the setting, and she longed for the days when once she sat among the violets, clean and untouched by shabby treatment.

She felt her mother's hand on her shoulder, and Tinette

then realized that she had been crying, actually sobbing. The weeping stopped abruptly, and the young woman turned to face her mother.

"I'm in the family way, Mama," she said, "but it will be all right."

When Ted ran in to wash for lunch, Tinette motioned for him to follow her. "Bring Franklin upstairs with you, I need help." She had the boys remove one of the cribs from the storeroom and place it in her own bedroom, hers and Earl's along the east wall. Then she sent them back for her old narrow bed and placed it at right angle to the crib.

"There," she said as the boys left the room. "There," she said as she stood looking down at her girlhood cot. "There I shall lie for the rest of my days."

Tinette was amazed that Earl questioned her reason for using her own bed. For a time he tried to plead innocence, argue helplessness, avow contrition. Tinette refused to answer any questions and refused to participate in any discussion. In controlled dignity she attempted to live on a higher plane keeping family life moving smoothly. She had regained her physical strength and the birth of another girl, whom she named Wanda, was as normal as the first three had been.

Strange to note, Earl's visits to Mary Pattison ceased completely. He even closed his office in town and seldom left the farm.

When news reached the farm that Mary Regina Pattison had been felled by a stroke which took her life, Earl sat in his room with downcast head. Was it in grief for the one, or was it in remorse for the other? He was named as the sole heir of the Pattison property, and Calvin helped him take care of its disposal. With money he gained from one large and two small sales, Earl Norris bought history books and sets of encyclopedias. He fixed himself a large desk next to his bed, and there he began to study and to compose a book of history as he saw it.

Eventually there was no crib left standing in the east bedroom. But one day before the birth of Wanda, Nancy had gone into that bedroom to pick up a scarf which Tinette needed in the dining room. The grandmother walked slowly and thoughtfully down the stairs to hand the scarf to Tinette. Seeing Calvin in the kitchen, she motioned for him to follow her silently. When they reached the top of the stairs, she led him into the bedroom and pointed to Tinette's old narrow bed. Calvin observed and nodded. Quietly they went back downstairs.

There was only one time that a word was spoken between them concerning the bed. Not long after their discovery, Calvin and Nancy stayed outside to watch the sunset, then as they turned toward the house Nancy said, "Tinette sent me after that scarf on purpose." Calvin agreed.

## CHAPTER TEN

## *The Farm*

All of the family gathered for Calvin's ninetieth birthday in 1922. In previous years various friends had been invited for birthday and anniversary events, but on this day Calvin asked for only the family. How long would he live, he wondered. Actually, he had not had too strong a start in life. He and his father had lived a hit-and-miss existence. Then both had received wounds in a pseudo part of the war, his father's being a mortal wound, and he himself had ailed from his wound ever since. For more than sixty years he had suffered more or less silently with his arm. Then, too, he had lived through privations and disasters, not to mention the hard work he had done under rugged conditions. But he was still alive.

More of a miracle to him was Nancy — old Nan, tiny and almost shriveled, yet still perambulating. It had been a wonderful day for him when he had been able to hire hands to help him so Nan could stay inside and do woman's work. That was because of Tinette. She had been born by then and she needed her mother. As it turned out, Nancy reveled in being indoors, but Tinette lived every minute that she could spare in the big outdoors. Usually she tagged Calvin, but she had her own activities around the farm, too.

As usual, when Calvin got to reminiscing, his thoughts soon got around to being chiefly of Tinette. A sweet baby, lovable child, outstanding young lady, and finally a stalwart and staunch woman. She was the best, not to be excelled and probably not equaled, as far as her father could see. Ted, of course, was the chief grandchild. Old Black Thunder had been one kind of chief. Ted was another.

It had been a great pleasure to his mother and grandparents the day Ted married Priscilla Katchall and brought her home to live. Priscilla was a grandchild of Henrietta and Julius Katchall, the O'Neil's closest friends, and to Calvin and Nancy Priscilla seemed like one of their very own. She had settled down with Ted in a back room upstairs, and Calvin had a large bathroom installed in place of the storeroom. The young couple brought joy and continuing harmony to the farm, and newcomers often noticed the close relationship between Tinette and Priscilla. Some thought they were mother and daughter instead of in-laws. Ted had assumed the work and management of the farm, while the women kept the big old house with its yards and gardens in shape. Part of their time, too, went to the care and enjoyment of Ted and Priscilla's offspring, four-year-old Marguerite. The little girl and Calvin together made an entity, a self-contained loyalty.

Ted was not the only grandchild who still lived under the old roof. The trilogy was there in its entirety. Ted, Sara, and Franklin, the equation that equaled one, had been reared so closely together that their cordon could hardly be broken. Even when Ted attached himself to the farm and to Calvin, it was as if he and the other two were only waiting for evening when the trio would recircle. It had been the same at school. No one could penetrate their tight bond.

Ted's affection found its object in Priscilla, but Franklin and Sara were not excluded and they continued to live at the farm. Franklin helped operate a store in town, and Sara clerked for him. At night they shared experiences and problems with Ted. Franklin loved Priscilla and Sara tried to. She hid her jealousy very well, but a prim stiffness was the result.

Tinette and Earl's youngest looked like Ted and like their grandfather. Her name was Wanda, and Wanda was like a jack-in-the-box. She was ready for any and every bit of fun and excitement. Black curls refused to pin up neatly, and she had promised herself to have her hair bobbed as soon as she thought everyone could take it. Wanda shied away from Ted because he always sided with Tinette in trying to tone her down. She thought of Sara and Franklin as old maids, and as often as she could she avoided them, too.

At the gathering, Wanda sat on the arm of Earl's chair and patted his shoulder. Marguerite would not leave Calvin, her dear great-granddaddy. Priscilla sat on a sofa with her mother-in-law, and Franklin perched on a stool beside Sara's chair as he usually did on such occasions. Also true to custom, Ted hovered near his grandmother, Nancy Hawkins O'Neil. Not one person, not even Wanda, wanted that day to end.

Ted Norris had no idea what disturbing news his brother Franklin was to worry the family with one morning in 1924. If he had, he probably would be copying the mannerism and expression of his Grandfather O'Neil by shaking his head from side to side and saying, "Some days it just don't pay to get up." As it was, he arose to face the day with an unsuspecting confidence.

"You going to get up, Priscilla?" he asked his wife who still lay asleep within the warmth of their big four-poster.

He stepped gingerly onto the cold carpet as he moved about the room to don his clothes for the day, all heavy apparel suitable to a farmer with outdoor chores awaiting.

"Priscilla!" he tried again. "Time's a-wasting, you

know." Not getting any response, and by this time being fully dressed, the black-haired, blue-eyed farmer walked in the protection of his thickly-soled shoes along the cold carpet by his bed.

"Now come on, Mrs. Norris," he blared out, pulling the covers back for emphasis.

"Quit, Ted," his wife whispered. "You'll awaken Marguerite yet." And so saying she reached for the covers and flung them tight over her round head.

"Well I'll be — let the work go — sleep, sleep, sleep," Ted was muttering as he turned from the bed and moved toward the door. With his hand on the doorknob he turned once more to look at his recalcitrant wife. "A blast wouldn't wake that little girl, and you know it."

"Quit once, Ted," his German wife repeated.

Ted Norris closed the door quietly upon his sleepy wife and slumbering daughter as he left their bedroom. Modernity had recently come to his grandfather's farm, and it was still such a novelty to Ted that entrance to the gleaming bathroom brought a certain amount of respectful reflection.

When he and Priscilla had moved into their present bedroom at the time of their marriage ten years earlier in 1914, this room that now possessed the luxurious plumbing had been a littered storeroom. For instance, in the very spot where Grandfather O'Neil's intended heir now sat pondering in retrospect, had been once the resting place for Grandmother O'Neil's trunk, the trunk that had seen the O'Neils through their early days at homesteading.

As Ted turned his gaze to the brightly glowing bulb above his head, he thought, "Grandfather was certainly a sly one when he had those Delco lights installed." He chuckled slightly as he remembered the old man's obvious delight in this latest gift to the family. "Of course, he did have Franklin attend to the actual business of it," he said

# THE FARM

to himself, "but to the rest of us it was a bang-up surprise."

Ted was always the first of the big family to arise. The responsibility of running his grandfather's fine farm had been his for a dozen years, but each day saw him awaken to his tasks with fresh interest.

Within the little village of Waterville, three-quarters of a mile north, Franklin Norris owned a half interest in the general department store, "Norris and Frye." Because it was really the only store that could boast the finest shoes for the whole family, as well as fresh produce for the table, Ted glibly dubbed it, "Ignore us and Fry."

The youngish farmer left the room and quietly stepped down the hall, barely able to suppress his morning enthusiasm in an urge to rap on his brother's door.

Ted reached the head of the stairs where the hall gave opening to two large bedrooms, in one of which slept his parents, Earl and Tinette Norris. Across the hall from them slept his two sisters, Sara and Wanda.

The ninety-two-year-old man who had homesteaded the property and enlarged the one-room frame house to its present enormity was sleeping peacefully as his eldest grandson passed the bedroom at the foot of the stairs. Not wanting to awaken his grandparents before their usual hour, Ted noiselessly made his way to the kitchen. He closed the door behind him and crossed the room to the old fireplace. The fire had been laid the night before, and as the young man brought it to life he wondered how many times the fireplace had gleamed with red-hot coals. An old Indian had had it ablaze the first night that Calvin and Nancy O'Neil had stepped into the house, sixty-one years previous. It had been often repaired and kept in more or less constant use ever since.

"I'll bet old Black Thunder never stirred around this early of a morning, at least not with much speed," Ted mused in remembrance of the oft-told tale of the Indian chief who once was companion and even unsolicited roommate of the O'Neils.

Satisfied with the blaze, Ted Norris went into the glassed-in milk porch where his heavy outdoor wraps were hanging. Then, ensconced in sheep-lined jacket, ear-covering cap and warm gloves, he left the house with lantern in hand, ready for his early morning chores.

The neighbors called Ted Norris "a real worker." Every day while it was still dark his lantern could be seen dimly shining as the firm hand that carried it went from chore to chore. A faint suggestion of the approaching dawn, along with the large copper lantern, shed enough light that young Mr. Norris could see his way out of the milk porch, down the rock-laid path of the side yard and out across the wide barn lot. By the time Ted had fed Pony and the pair of grays, the pinkish rays that precede a bright sunrise were only beginning to appear. But between the feeding of Boss and her calf, which at present were being accommodated in the corral, the sun had made its first appearance. And as Farmer Norris went back to the house, lantern still swinging in his hand, he could look straight into the bright ball that seemed to be shooting up now above the fields across the road.

That road, though community property, was like a prized possession. It was a definite line that separated that piece of land from that which belonged to Sam Stanton, and separated it in a neat, orderly way that was right to the liking of Ted and his way of doing things. Now take the other boundary markings of the farm. That neighbor of his to the south, Loy Parker. That lazy son-of-a-gun had never been known to lay a decent fence. If the stones gave the appearance of a well-laid fence, it was because of a lopsided arrangement where one man does the work while the other keeps up a line of gab. Otherwise the farm was marked off more vaguely. The northernmost section was used for pastureland and came to rather an abrupt end where some jagged rocks crossed the property. Ted was unwilling to mark the rest of the land in hopes of acquiring the section which would change the shape of the farm from an inverted "L" to a complete rectangle.

This was not so much a wish of his own as a desire transplanted to his mind by his grandfather. At intervals frequently interspersed among the span of Ted's thirty-two years, Calvin had related to his grandson, as well as to Ted's brother and two sisters, the story of his wish to own the property, and the thwarting of his every attempt to do so. It was correctly called a piece of unfinished business. Anyway, back to the east there lay the straight road that gave marked definition to that side of the O'Neil farm.

As Ted finished the daybreak chores he noticed that only one of the rooms upstairs was alight with the Delco lights. It was the room to the south, the former storeroom lately transformed into a shining example of modernity and comfort. As he entered the milk porch he could also see a strong light glowing under the door which led to the kitchen. From inside came the accustomed sound of breakfast being prepared by his wife and his mother.

With a few quick movements he extinguished the light in the lantern, and in the semi-darkness removed his heavy winter wraps. He entered the kitchen with a fast, jumping stride and crossed to the fireplace to warm his stiff hands.

"Um-um, that smells good. Is it about ready?" He asked no one in particular.

Priscilla gave Ted a slow smile. Then Tinette, who had been forming a question just as Ted spoke, asked, "Is it awfully cold out this morning, Ted? I nearly froze while dressing, but those north rooms are always cold of a morning. Is it bad outside?"

Ted muttered something about the cold as he turned his back to the fire and stood with his hands resting on the small of his back. His head of black hair, neither straight nor curly but only unruly, accented his bright blue eyes and gave him a very youthful appearance. Ted wore his habitual expression of well-being as his brother entered from the dark sitting room.

Franklin Norris never did have the happy, carefree look of his brother, but on this morning he had a more earnest expression than usual. His mother noticed it as he gave her a good-morning pat and went on to the stove to pour himself a cup of coffee. In all the thirty years that Ted had had Franklin as a brother, he had never seen him up so early. He greeted him good-naturedly.

"What's keeping you up in the middle of the night? Or are you just getting in from that party?"

"You got up to tell me all about Alice's party, didn't you, Franklin?" Priscilla asked.

Franklin could not help relaxing into a smile at this subtle bit of curiosity from his sister-in-law.

"Hardly, Sis," he said. "That Alice Hall and her parties are going to be the death of me yet." Turning to Ted he continued, "But I did get one bit of information that I think you and Mother should know about."

Tinette paused in her sausage frying and Ted pulled chairs up to the table for himself and Franklin.

"Remember Lon Baker?" Franklin asked. "He was here for Grandma's diamond wedding celebration last spring. Came as a guest of Judge Ellis."

Ted puzzled over the name a moment, "Lon Baker. No, I can't place him."

"Oh, you know him," Franklin insisted. "Think back. At the celebration he let it slip that he was a bachelor and Miss Evie began following him around. Don't you remember, Ted?"

"I know who you mean," Tinette said. "He even came back later in the summer to get the name of Mama's yellow rose bush. You should remember that, Ted. But what about him, Franklin?"

"Well, he is a lawyer for Fred Ross and I ran across him at the party last night."

"You think he's after Grandma's yellow roses?" Ted teased.

"Let me fill my coffee cup and I'll tell you. It's not something I want to say too loudly." He poured coffee all the way around and then went on. "Baker says that Bill Walton was in his office yesterday about that property next to ours. It seems Walton has a buyer and intends to sell."

For a moment everything was quiet in the room. Only the soft sounds of the breakfast frying continued as if nothing had happened. Ted was the first to speak. "But he has had a buyer for it ever since he's owned it. Grandfather said he no sooner learned who had got the property than he gave him an offer."

"As far as I know," said Franklin glancing at his mother, "Bill never once considered letting Grandpa have it."

Tinette quickly busied herself with the toast.

Ted's blue eyes took on the color of cold slate and filled with disdain as he said, "Bill Walton! Sneaking around to sell that land behind our backs. The snake!"

"That's what he is, a snake. A slimy snake," Franklin spoke quietly but not quite resignedly. "Grandpa has been looking to own that property for over half a century."

"Many is the time I have stood with Grandfather at the edge of our land, and he would tell again and again of trying to buy that little section," Ted said.

"He wants to complete the farm, own the entire piece. Is that it?" Priscilla asked.

"It yields well, too, Sis," Franklin said. "Bill would never let us have it."

"Don't forget that we have no access to the big road unless Grandfather could get that land away from Bill somehow. There seems to be something in the situation that Grandfather has never come through with. At least not to me. I've seen him shake his head over it like he does when he wants to say something, and then stop short. Has he ever told you, Mom?"

Tinette's expression was vaguely a knowing one but her answer was evasive. "I can just see Papa shaking his head over it, running his fingers through those tight black curls."

"They're white now, Mother," laughed Priscilla.

"I always see them black," Tinette reflected. "I've known him for a long while."

Franklin winked at Ted and said, "Over fifty years, isn't it?"

Tinette paid no attention and went on speaking. "And in the early part of those years Papa was too poor to buy the land if it had been for sale."

"Well, Mom," Ted corrected, "if I remember what he told me, at first he did have the money to pay for it but the owner wasn't around or wouldn't sell or something."

"Then he was too poor to buy," Franklin added in summary, "and for a long time he has been in the same fix he was originally."

"The owner just won't sell. At least not to Grandfather," Priscilla said with a tone of finality mixed with incredulity.

"That's right, Sis," Franklin said, "and now Bill has decided to sell and Grandpa's too old to try to fight for it."

Ted's eyes again slated into cold fury as he said, "This could kill Grandfather, as old as he is, this kind of worry."

Franklin nodded and said gravely, "That is what I've been thinking. If he hears that Walton has decided to sell, nothing will keep him from an active part in it."

It was now fully daylight and voices could be heard from upstairs. Then someone was heard crossing the sitting room, and the boys' mother whispered, "Papa must never know."

Grandfather O'Neil had not always been a ninety-two-year-old man from whom worrisome secrets had to be kept. His daughter easily remembered him as a person so commanding that all decisions rested with his judgment. Just ten years previously, heartily approving of Ted's marriage to Priscilla Katchall, he had sat at the big library table in the sitting room one evening and had drawn up plans for another enlargement of the house. At that time he had still risen at dawn and had supervised every bit of the construction.

Yet Tinette realized in her retrospection that soon after completion of the work, her father had rapidly let go of the responsibility for running the farm. By the time his great-grandchild had entered the world, he had become content to stay indoors most of the time. He began to rise later than ever before and even took to letting Tinette take breakfast to himself and Nancy. Later he would walk out of the large downstairs bedroom, through the fine center hall, and back to his chair in the sitting room. This had become a signal for Priscilla to lay the baby in the cradle opposite him. They spent long periods of the day silently getting acquainted. There was a thorough bond between them despite the fact that words were few between them, even after the child was chattering gaily with other members of her family.

Calvin O'Neil had realized from the time that Ted was a small boy that he would be the grandchild to carry on the work of the farm. This may have been the reason Nancy had thought that her husband treated Ted "a mite more tender" than he dealt with the others. Anyway, Ted was "his," and to watch the old man with his little

Marguerite, one would think that she would carry on her daddy's work to preserve the farm.

But it was true, Tinette mused, that Grandfather was not always an old man from whom secrets could be kept. She was saddened to realize how very old her dear father really was.

As Tinette had been away in her thoughts, her daughter Sara had entered the room. She was standing between her two brothers, very prim and proper and looking for all the world like the third member of their triplicity. Each one had one essential quality that marked him from the others. Ted, the oldest, almost always sported a good-natured grin, while Franklin, the youngest, carried a look of worry on his face. Sara was undeniably a self-centered spinster. Of course, she was thirty-one, but Sara had been an old maid even as a first grader.

Sara asked Franklin why he was up so early, then she almost managed a smile to Priscilla. Had it not been for the sweet guilelessness of Priscilla's nature, Sara would have found it within herself to be resentful of the fondness her family showed her. But the sister-in-law had been able to level off Sara's arrogance with her own humility. Thus it was with congeniality that the five of them sat down to a delicious and more than ample breakfast.

Priscilla Norris, born into the thoroughly German Katchall family, and her mother-in-law who was of Irish stock, looked surprisingly alike. Both possessed fat bodies and plump, placid faces. Priscilla's light brown hair and Tinette's of gray wound around their heads in like-style braids. Even their blue eyes shone large and bright, though Tinette's were a deeper blue.

"Franklin, please pass the honey around," Priscilla said. "Ted, better have some more hot toast," she added.

"You know, I believe it is worth getting up a half hour early just to eat more leisurely." Sara said.

# THE FARM

"Worth it just to eat more, you mean," answered Ted.

"Well, for once you can take your time," Tinette said sympathetically and then returned to the question she had asked earlier, "Ted, is it as cold today as it was yesterday?"

"Yes, Mother. I told you it was every bit as cold. Or I thought I did."

Sara asked who had been at Alice Hall's party, and before he thought, Franklin named Lon Baker. A strange quiet fell upon her family but at the moment she was too interested in getting her second cup of coffee just the right color to notice any effect a name would have on anyone. Almost immediately Priscilla jumped up, saying she had better awaken Marguerite and Wanda. Her mother-in-law answered with a piece of toast still in her mouth, "All right. And tell Wanda that I said to dress warmly. She will know what I have in mind." Tinette gave a little laugh because it was in the days of long underwear and high school girls were beginning to rebel against it.

As the short, plump Priscilla went out of the kitchen she received a smile from Ted and returned it. She stayed in the sitting room long enough to pull back the heavy drapes that had been keeping the dawn's light from entering through the bay window. Then she hurried on through the hall, passing quietly the room where the O'Neils lay sleeping, and on up the stairs.

Wanda was already about to awaken because of the cold that had crept in when Sara had left the covers thrown back. The seventeen-year-old Wanda had only one resemblance to her mother's family, and that was the black, curly hair she had definitely inherited from her Grandfather O'Neil. Priscilla bent over the bed, ran her hand through the short curls, and said, "Wanda. Jump right up." And Wanda did. The first and last born in the Norris family had the same habit of getting up early and starting the day at full speed.

"Oh, yes," Priscilla said as she paused at the door. "Your mother said to dress as you did yesterday."

Not wanting to hear Wanda's groan, she hurried along through the hall to her own bedroom to awaken Marguerite. Priscilla helped the little six-year-old dress quickly and then sent her down to the kitchen. No one was there but Tinette, and Marguerite told her, "I wish I could stay at home with you, Grandmother. All day long."

Little Miss Marguerite Norris had the raven black hair of Calvin O'Neil, but it was noticeably without curl. Her neatly combed hair, separated by a center part, was held snugly to the base of her head by a small bow of ribbon behind each ear. The shiny hair dropped over each shoulder in tight braids. Her fine, sweetly curved lips formed a replica of her grandmother's smile.

"This is fun to be alone, isn't it?" The little girl followed her question with an ingratiating wink, one similar to her father's. She began to eat her ample breakfast when her Aunt Wanda entered the kitchen and perched herself on the edge of a stool, reluctantly drinking a cup of cocoa and refusing all other food. Sara walked into the room and brusquely said a good morning. There was no response from either girl, and after Priscilla had run in to help Marguerite don some warm wraps, the three younger females went through the milk porch to the side yard to catch their ride.

Sara was Franklin's efficient clerk, and she never laughed when Ted referred to the store as "Ignorus and Fry." She and Franklin rode to town together, taking Wanda and Marguerite to school on their way. Wanda went to another exciting day as the most popular girl in the senior class, while their little niece went to a long, trying day as a first grader in a strict teacher's class.

As soon as Ted had finished breakfast he had walked with Franklin to the lot where they kept their cars. Neither had owned a car until the previous fall. At that time Ted bought a secondhand Chevrolet touring car and had built a garage large enough for two automobiles with a little space left over. The car that Franklin had purchased was a softly purring Buick.

As the two brothers walked across the yard, Ted asked, "What is our best move, Franklin?"

Franklin had his reply ready. "We'll have to go down and talk to Bill."

"Think he'll listen?" Ted asked.

"He's got to."

"I'm willing to give it a try," Ted said. "Here come the girls. So shall I meet you at Walton's place?"

"One o'clock all right?"

Ted nodded and helped the ladies into the car. He covered their legs with a robe and waved them goodbye.

At noon Ted told his mother what he and Franklin were planning to do. When Tinette saw him drive out of the driveway she could only shake her head. "It will never work, dear boys," she said softly. During the morning hours she had done her kitchen work with her mind on other days. She thought back to when she had first seen Bill Walton and to the time when she had been completely sickened by him.

She took time to have a visit with her parents. She let each of them tell her about their restlessness in sleep and how they could feel the cold even with the warm fire that Ted kept going in their room. She looked at her father and wanted to hold him so he could not ever hear that his dream of having the coveted land was to be unrealized. When Earl took his breakfast, only a cup of coffee, Tinette recalled how it was that she finally consented to his proposal of marriage. She speedily turned that picture out of her mind. It had been years since she had thought of Weston O'Neil. The name still gave her a pang, and she went into the cold parlor to put her feelings on a different plane. Later she glanced in at the old folks and they were having a pleasant game of dominoes, one of which tired the grandfather.

Calvin O'Neil awakened from his nap still feeling sluggish and as though he could relish a bite of cheese. He found just what he wanted on the kitchen table and took his refreshment to the sitting room. He sat in his rocker, leaning forward to look out the bay window at his tall pine tree. All at once he saw Ted come driving into the driveway. Now Ted never left during the day so this was a puzzle. The door opened and Calvin turned to ask Ted where he had been, but instead of Ted it was Marguerite who entered.

"Well! Did your dad pick you up at school today?"

"No, I was walking by old Walton's place and there was Daddy ready to leave. So I got a ride home."

"You don't mean that your dad had business at Walton's!"

"I don't know. But, Grandfather. Why does old Walton want to sell to someone with our name, only it isn't us?"

Suddenly she put a finger to her lips and tiptoed out of the room. Ted and Priscilla were in the kitchen talking quietly.

Calvin strained every muscle he had to lean over and listen to what they were saying. Among a few other words, one came out clearly. It was "Weston O'Neil." The door opened and Calvin pretended to be asleep. When he knew he was alone again he went back to his room. He drew his high wing-back chair to the north window and sat down.

Weston O'Neil. So that was it. Bill was going to do a little spite job and sell the property to a different O'Neil. But Walton did not know the story. He didn't even know there was a story. Nancy had gone upstairs with Tinette, and Ted had helped her down. As the two entered the room, Calvin pushed back his chair and stood up.

"Ted, I want you to get my coat and cap. I want you to take me down the road apiece."

"Why sure, Grandfather. But it's awful cold out there, and not far from dark, too."

Calvin did not answer so Ted helped him put on some warm wraps. They walked slowly to the garage and Calvin puffed as he got into the car.

"Where to, Grandfather?"

"Down to Walton's, I reckon, Ted."

Ted was taken completely by surprise. Turning events over in his mind, he felt sure that his grandfather could not possibly have heard the family talking about the matter. Yet why else would he be making the trip?

The half-mile was soon covered and Ted drove into Bill Walton's yard. He jumped out of the car to help the old man. The ground was unfamiliar to Calvin, and in spite of the strong arm of his grandson he pecked his way slowly. As they reached the porch the door opened and Walton faced them, scowling. Calvin asked his grandson to wait outside while he went in.

Ted Norris kept asking himself how Grandfather found out; then he realized that the old man had been in the adjoining room when he and Priscilla had been talking about it. That was it. It had to be. The lights of a car shone up the hill. It was the Buick, taking Franklin and Sara home. Neither saw Ted's car in the yard. If Grandfather didn't hurry, everyone would know that they were gone. He recalled the angry session of the afternoon. Bill simply would — His thoughts were caught short by hearing his grandfather's voice calling to him. The door was open and there stood the two men, Calvin erect and calm, Bill sheepish with eyes downcast.

"Ted, come shake hands with Bill. We've signed up."

"What have you signed, Grandfather?"

"Why, my check, of course."

Ted was alarmed and asked, "What have you signed, Bill?"

"The deed," he said. "I had everything here so I could, uh — well, the other fellow was to be here in a day or two." He accepted Ted's outstretched hand.

Then the old man explained that the deal was settled, but that it was to be in Ted's name, so Ted would have to meet Bill in Marysville for the legal end of it. Ted saw that his grandfather was tired so he helped him to the car and let all further questions go for the time being.

All the way home, Calvin sat in silence, too tired to talk. Tuckered out, as he said. He put his head back against the seat and relaxed. Arriving home he told Tinette to give him a few minutes before supper. Then he went to the large front bedroom and closed the door behind him. He sat down in the wing-back chair opposite Nancy, giving her a wink as promise of a secret.

Speaking softly he said, "Bill sold me the eighty acres."

She fairly gasped, "Why, Calvin? Tell me."

"Know who he was aimin' to sell it to?"

Nancy shook her head.

He leaned close to her and whispered, "To Weston O'Neil."

His wife sank back in her chair as if the name scared her. "Weston O'Neil," she repeated in a low voice. Then she straightened up and asked, "Did you tell him?"

Calvin nodded his head. "Yes, I did. I figgered Bill wouldn't be apt to make it possible for Tinette to be near the only man she ever really loved."

Both of them remained silent for a long moment. Then Calvin added, "Seemed like he was even feelin' a mite sorry for her. He said he wanted to do it as sort of token to

Tinette. We put it in Ted's name."

"I guess he realized that Tinette has had her heartaches, too," Nancy concluded.

Together they rose, and taking each other's arm they walked out into the hall, and back to the family they loved.

## THE END

# Share the Warmth of the Capper Fireside Library!

**Other Capper Fireside Library Titles Currently Available:**

These Lonesome Hills ◆ Letha Boyer
Home in the Hills ◆ Letha Boyer
Born Tall ◆ Garnet Tien
The Turning Wheel ◆ Garnet Tien
Carpenter's Cabin ◆ Cleoral Lovell
The Farm ◆ LaNelle Dickinson Kearney
The Family ◆ LaNelle Dickinson Kearney

——————— ◆ ———————

For more information about Capper Press titles
or to place an order, please call:
(Toll Free) 1-800-777-7171, extension 107,
or (913) 295-1107.